Raoul Vyumvuhore

Hydratation et propriétés mécaniques du Stratum Corneum

Raoul Vyumvuhore

Hydratation et propriétés mécaniques du Stratum Corneum

Validation des descripteurs de l'hydratation et des propriétés mécaniques du Stratum Corneum ex vivo et in vivo

Presses Académiques Francophones

Impressum / Mentions légales
Bibliografische Information der Deutschen Nationalbibliothek: Die Deutsche Nationalbibliothek verzeichnet diese Publikation in der Deutschen Nationalbibliografie; detaillierte bibliografische Daten sind im Internet über http://dnb.d-nb.de abrufbar.

Information bibliographique publiée par la Deutsche Nationalbibliothek: La Deutsche Nationalbibliothek inscrit cette publication à la Deutsche Nationalbibliografie; des données bibliographiques détaillées sont disponibles sur internet à l'adresse http://dnb.d-nb.de.

Coverbild / Photo de couverture: www.ingimage.com

Verlag / Editeur:
Presses Académiques Francophones
ist ein Imprint der / est une marque déposée de
OmniScriptum GmbH & Co. KG
Heinrich-Böcking-Str. 6-8, 66121 Saarbrücken, Deutschland / Allemagne
Email: info@presses-academiques.com

Herstellung: siehe letzte Seite /
Impression: voir la dernière page
ISBN: 978-3-8381-7342-9

Zugl. / Agréé par: Chatenay-Malabry, Université Paris Sud, 2013

Raoul VYUMVUHORE

VALIDATION ET IMPLEMENTATION DES DESCRIPTEURS DE L'HYDRATATION ET DES PROPRIETES MECANIQUES DU STRATUM CORNEUM *EX VIVO* ET *IN VIVO*

Quand les spectroscopies vibrationnelles
rencontrent la peau...

A ma femme et à mon fils qui m'ont soutenu pendant mes travaux

A mes frères et sœurs

A mes parents

1

REMERCIEMENTS

Je remercie de tout mon cœur ma directrice de thèse : Pr. Arlette Baillet-Guffroy de m'avoir suivi pendant ces cinq dernières années. Cette thèse n'aurait jamais vu le jour sans votre confiance et votre sympathie. Merci infiniment pour votre grande générosité, votre enthousiasme ainsi que pour vos précieux conseils qui ont fait progresser ce travail. Au-delà de vos qualités scientifiques, je vous remercie également pour vos qualités humaines. Vous avez toujours cru en moi et je vous en remercie. Je n'oublierai jamais combien vous vous êtes battue pour moi tout au long de cette thèse sur différents volets. Vous avez un grand cœur. Je vous exprime donc toute ma gratitude et vous dis mille mercis.

Ce travail ne serait rien sans Dr. Ali Tfayli, mon encadrant. Tu as été plus qu'un encadrant. Tes immenses qualités scientifiques, tes visions stratégiques, ainsi que tes conseils pratiques ont fait énormément progresser ce travail. Ton aide a été précieux du début à la fin, j'admire ta clairvoyance et la rapidité de ta réflexion. Cette thèse s'est déroulée dans la joie et la bonne humeur grâce à ton écoute, ton enthousiasme, ta patience sans fin... et tant d'autres qualités! Je te remercie infiniment pour ta disponibilité ; aucun doctorant ne peut rêver mieux. Ne changes rien, tu ne feras que des heureux. J'ai beaucoup appris en travaillant à tes côtés.

Mes remerciements s'adressent au Pr. Pierre Chaminade, directeur du Groupe de Chimie Analytique de Paris Sud, pour m'avoir accueilli au sein du laboratoire, et pour ses discussions pendant les pauses aussi bien enrichissantes que divertissantes.

Je remercie évidement le Centre Européen de Recherche sur la Peau et les Epithéliums de Revêtement (CERPER) pour avoir financé ma thèse.

Je tiens à remercier Pr. Philippe Humbert de l'Université de Franche-Comté de l'honneur que vous m'avez fait en acceptant d'être rapporteur de mon travail de thèse malgré votre emploi du temps très chargé.

Je remercie Igor Chourpa professeur à l'Université Francois-Rabelais de Tours, rapporteur de cette thèse, pour l'intérêt qu'il a manifesté pour mes travaux de recherche. Merci d'avoir alloué votre temps précieux à l'évaluation de mon travail.

Je suis très sensible à l'honneur que m'a fait l'ensemble des membres du jury de lire et de juger ce travail : Pr. Jean Doucet, Dr. Hélène Duplan, Pr. Stéphanie Briançon, Dr. Alexandre Delalleau, Pr. Michel Manfait, Pr. Reinhold Dauskardt, Pr. Pierre Chaminade.

Je remercie chaleureusement Dr. Hélène Duplan des laboratoires Pierre Fabre pour votre collaboration et pour votre contribution scientifique. C'était très agréable de vous retrouver lors des différentes réunions de travail. Merci de m'avoir aidé à valoriser mes travaux à travers les financements des congrès et d'avoir défendu la publication de nos travaux. Merci beaucoup pour votre soutien et votre générosité.

Je remercie le laboratoire Silab, en particulier Maud Le Guillou, de m'avoir ouvert l'accès aux mesures *in vivo* de la peau.

Un immense merci à Dr. Alexandre Delalleau, qui m'a fait bénéficier de ses compétences et de ses enrichissants conseils. Tu m'as initié aux mesures mécaniques avec beaucoup de patience et de gentillesse. Merci pour tes suggestions et remarques toujours pertinentes.

Merci au Pr. Reinhold Dauskardt pour les diverses discussions scientifiques qui ont beaucoup contribuées à l'amélioration de la qualité de ce travail. Grâce à vous j'ai compris que c'est les détails qui font la différence. Je vous remercie également pour l'acceuil dans vos laboratoires à Stanford.

Une thèse ne serait pas vraiment une thèse sans les personnes qui aident et conseillent. Je souhaite dire:

Un grand merci au Pr. Michel Manfait et au Pr. Olivier Piot pour m'avoir permis de réaliser mes expériences dans leur laboratoire pendant mes deux premières années de thèse.

Merci à Valérie Untereiner, pour m'avoir facilité l'accès aux appareils

Merci à Krysta Biniek pour sa contribution dans la partie biomécanique de cette thèse et pour son acceuil à l'Université Stanford.

Merci à mes stagiaires Marie Aure Hallay, Hai Yen TA, Larry BUFFLE pour leur contribution à la réalisation de ce travail

Un merci tout particulier à Rabei Mohamedi pour son efficacité à passer mes commandes, pour son aide et sa gentillesse et à Sonia Abreux.

Merci à Johanna Saunier, pour les expériences sur la DSC et l'ATG.

Merci à l'ensemble du laboratoire de Chimie analytique de Chatenay-Malabry.

Je tiens à remercier également la famille Alex Adhémar Ndikumana pour son aide et son soutien dès mon arrivée ici.

Je n'oublie pas aussi de remercier tous mes amis avec qui j'ai passé de bons moments.

Je remercie tout naturellement mes parents, mes frères et mes soeurs, tout en faisant un petit coucou à Happy et Titi.

Enfin, de tout mon coeur, je remercie ma petite femme Katia de son amour, de m'avoir fait confiance et de m'avoir soutenu tout au long de mon parcours.

Je dédie ce travail à mon fils Enzo qui représente pour moi une motivation infaillible dans tout ce que je fais.

RÉSUMÉ

La peau est l'organe le plus grand du corps humain et représente ~10% de la masse corporelle. La bonne qualité de son état et de ses fonctionnalités est primordiale pour la santé d'un individu. La sécheresse cutanée constitue un phénomène commun dans différents dysfonctionnements physiopathologiques. Grâce à ses propriétés de protection de l'organisme vis-à-vis de son environnement, le *stratum corneum* (SC) est considéré comme le principal élément contrôlant l'hydratation. Ce travail de thèse, associant des développements techniques et méthodologiques, a conduit à la mise en évidence par microspectroscopie Raman confocale, des mécanismes moléculaires impliqués dans les phénomènes de sécheresse cutanée. Le lien moléculaire entre hydratation et stress mécanique du SC *ex vivo* est décrit de manière approfondie impliquant lipides et protéines tissulaires. Ces travaux ont également porté sur la caractérisation des modifications supramoléculaires responsables des déformations du SC sous stress mécanique. En parallèle, ce travail illustre l'intérêt des spectroscopies vibrationnelles comme outil d'évaluation des mécanismes d'action des produits hydratants.

Le caractère non-invasif de la spectroscopie Raman a permis d'exploiter les fortes potentialités de cette technique en transposant *in vivo* l'utilisation des descripteurs spectraux obtenus *ex vivo*. Ainsi, nous avons développé une approche *in vivo* couplant la spectroscopie Raman et la méthode des moindres carrés partiels (PLS) pour la quantification indirecte de différents paramètres physico-chimiques et fonctionnels du SC y compris les lipides et l'eau conduisant à une caractérisation globale du statut physiopathologique du SC.

Mots clés : spectroscopies Raman et infrarouge, hydratation, propriétés mécaniques, *stratum corneum*, produits hydratants

ABSTRACT

The skin is the largest organ of the human body, accounting for ~10% of the body weight. The quality of its state and functionality is essential for the human health. Dry skin is a common phenomenon in various physiopathological dysfunctions. The uppermost layer of the skin, the *stratum corneum* assumes the first barrier between organism and environment, it is thus considered as the main element controlling skin hydration. This work, combining technical and methodological developments, led to highlight the molecular mechanisms involved in skin dryness phenomena by confocal Raman microspectroscopy. The molecular link between hydration and mechanical stress of SC *ex vivo* is described in detail involving lipids and proteins. This work has also focused on the characterization of supramolecular changes related to the deformations of the SC under mechanical stress. In parallel, this work illustrates the effectiveness of vibrational spectroscopy for evaluation of moisturizers mechanisms of action.

The non-invasive nature of Raman spectroscopy allowed exploiting the high potential of this technique by transposing spectral descriptors obtained *ex vivo* to *in vivo*. Thus, we developed an *in vivo* approach coupling Raman spectroscopy and partial least squares method (PLS) for indirect quantification of different physico-chemical and functional parameters including the SC lipids and water leading to an overall characterization of the SC physiopathological status.

Keywords: infrared and Raman spectroscopy, hydration, mechanical properties, *stratum corneum*, skin moisturizers

Ces travaux ont fait l'objet des productions scientifiques suivantes :

Articles de revues

Vyumvuhore R, Tfayli A, Duplan H, Delalleau A, Manfait M, Baillet-Guffroy A: Effects of atmospheric relative humidity on *stratum corneum* structure at the molecular level: ex vivo Raman spectroscopy analysis. **Analyst 2013**; 138:4103-4111

Vyumvuhore R, Tfayli A, Duplan H, Delalleau A, Manfait M, Baillet-Guffroy A: Raman spectroscopy: a tool for biomechanical characterization of *stratum corneum*. **Journal of Raman Spectroscopy 2013**; 44:1077-1083

Tfayli A., Jamal D., **Vyumvuhore R.**, Manfait M., Baillet-Guffroy A.: Effect of the hydration on the *stratum corneum* lipids barrier function: Raman analysis on ceramides 2, III and 5. **Analyst 2013;** 138:6582-6588

Vyumvuhore R., Tfayli A., Baillet-Guffroy A.: Vibrational spectroscopy and classical least square analysis, a new approach for determination of skin moisturizing agents' mechanisms. **Skin Research and Technology 2013 Nov 21. doi: 10.1111/srt.12117**

Vyumvuhore R., Tfayli A., Biniek K., Duplan H., Delalleau A., Manfait M., Dauskardt R., Baillet-Guffroy A.: The relationship between water loss, mechanical stress, and molecular structure of human *stratum corneum ex vivo*. **Journal of Biophotonics 2014 Jan 21. doi: 10.1002/jbio.201300169.**

Vyumvuhore R., Tfayli A., Piot O., Guichard N., Le Guillou M., Manfait M., Baillet-Guffroy A.: Raman spectroscopy: *in vivo* Quick Response "QR" code of skin physiological status. **Journal of Biomedical Optics 2014 Nov; 19(11):111603. doi: 10.1117/1.JBO.19.11.111603.**

Communications orales

Vyumvuhore R., Tfayli A., Biniek K., Duplan H., Delalleau A., Manfait M., Dauskardt R., Baillet-Guffroy A.: Skin dryness modifies the lipid barrier state and protein structure. *JED Chimie de PARIS SUD*, September 2013, Orsay, France

Vyumvuhore R., Tfayli A., Biniek K., Duplan H., Delalleau A., Manfait M., Dauskardt R., Baillet-Guffroy A.: Skin hydration and mechanical stress mechanism: molecular aspects of the barrier function and protein structure. *European Conference on the Spectroscopy of Biological Molecules (ECSBM)*, August 2013, Oxford, UK

Vyumvuhore R., Tfayli A., Duplan H., Delalleau A., Manfait M., Baillet-Guffroy A.: Molecular mechanism of *stratum corneum* elasticity: investigation by Raman spectroscopy. *Congress of International society for biophysics and imaging of the skin (ISBS)*, November 2012, Copenhagen, Denmark

Tfayli A., **Vyumvuhore R.**, Manfait M., Baillet-Guffroy A.:Characterization of skin barrier function using Raman spectroscopy: new highlights in skin hydration. *Congress of SPEC*, November 2012, Chiang Mai, Thailand

Vyumvuhore R., Tfayli A., Duplan H., Delalleau A., Manfait M., Baillet-Guffroy A.: Effects of atmospheric relative humidity on *stratum corneum* lipids and proteins structures: Raman spectroscopy analysis. *Congress of International Society For Stratum Corneum Research*, Stratum Corneum VII, September 2012, Cardiff, UK

Communications par poster

Vyumvuhore R., Tfayli A., Manfait M., Baillet-Guffroy A.: Vibrational spectroscopy and classical least square analysis for quantification of moisturizing agents' effects on the *stratum corneum* structure. *Congress of European Biophysical Societies' Association (EBSA)*, July 2013, Lisbon, Portugal

Vyumvuhore R., Tfayli A., Duplan H., Delalleau A., Manfait M., Baillet-Guffroy A.: Elastic behavior of *stratum corneum* : molecular investigation by Raman spectroscopy. *Congress of International Society for Biophysics and imaging of the Skin (ISBS)*, November 2012, Copenhagen, Denmark

Vyumvuhore R., Tfayli A., Baillet-Guffroy A.: Validation and implementation of skin hydration descriptors *ex vivo*. *JED de PARIS SUD*, September 2011, Orsay, France

SOMMAIRE

REMERCIEMENTS..2

RESUME 5

ABSTRACT 6

PRODUCTIONS SCIENTIFIQUES..7

SOMMAIRE..9

LISTE DES ABREVIATIONS 12

LISTE DES FIGURES 13

LISTE DES TABLEAUX.. 15

INTRODUCTION GENERALE... 16

ETAT DE L'ART.. 20

Chap I. *Stratum corneum*: structure, fonctions et propriétés................. 21

   I.1. Structure du *stratum corneum*.. 22

   I.2. Hydratation cutanée et fonction barrière... 27

      I.2.1 Régulation de L'hydratation du *stratum corneum*......................27

         I.2.1.1 Absorption d'eau dans le *stratum corneum*......................27

         I.2.1.2 Barrière à la diffusion de l'eau.. 28

         I.2.1.3 Système de contrôle et de réponse adaptative 29

      I.2.2 Caractéristiques d'une peau sèche.. 29

      I.2.3 Gestion de la peau sèche: Produits hydratants.......................... 30

      I.2.4 Méthodes d'évaluation objective de la peau sèche 31

         I.2.4.1 La perte insensible en eau.. 32

         I.2.4.2 Méthodes électriques de mesure de l'hydratation du *stratum corneum*............ 34

         I.2.4.3 Spectroscopie Raman confocale et hydratation cutanée................. 36

         I.2.4.4 Mesure des propriétés mécaniques du *stratum corneum* 37

Chap II. Principales techniques analytiques mises en œuvre dans l'étude tissulaire du *stratum corneum*... 43

   II.1. Les spectroscopies vibrationelles : principe et instrumentation 44

      II.1.1 Notions de base de la spectroscopie Raman.......................................46

II.1.1.1 Principe de l'effet Raman 46

II.1.1.2 Spectre Raman 49

II.1.1.3 Appareillage Raman 51

II.1.1.4 Micro-sonde Raman 54

II.1.1.1 Prétraitement des spectres Raman 55

II.1.2 Notions de base de la spectroscopie infrarouge 57

II.1.2.1 Principe de l'absorption infrarouge 57

II.1.2.2 Appareillage infrarouge 59

II.1.2.3 La Réflexion Totale Atténuée (ATR) : 60

II.2. Instruments de mesures mécaniques du *stratum corneum* 62

II.2.1 Mesures de courbure du substrat : Stress bi-axial du *stratum corneum* 62

II.3. Analyses statistiques multivariees 64

II.3.1 Analyses statistiques descriptives 65

II.3.1.1 Analyse en composantes principales (ACP) 65

II.3.2 Analyses explicatives et descriptives 66

II.3.2.1 Régression: moindres carrés partiels (PLS) 66

II.3.2.2 Analyse discriminante par moindres carrés partiels (PLS-DA) 66

II.4. Humidité relative 68

Chap III. OPTIMISATION DES CONDITIONS D'ANALYSE et DES PARAMETRES D'ACQUISITIONS **72**

III.1. Optimisation de l'environnement d'analyse 73

III.1.1 Contrôle de l'humidité autour du microscope du Raman 73

III.1.1.1 Controleur d'Humidité relative 01 (RHC-01) 73

III.1.1.2 Controleur d'Humidité relative 02 (RHC-02) 75

III.2. Optimisation des paramètres d'acquisitions 78

III.2.1 Microspectromètre LabRam HR 79

III.2.2 Microspectromètre LabRam 81

III.3. Étude de la variabilité du signal spectral Raman 83

Chap IV. Caractérisation de l'hydratation du *stratum corneum* **85**

IV.1.   Préambule 86

IV.1.1   Quantification de l'eau du *stratum corneum* 86

IV.1.2   L'effet de l'humidité relative sur la Structure des lipides et des protéines du
stratum corneum 88

Article 1 : Effects of atmospheric relative humidity on *stratum corneum* structure at
the molecular level : ex vivo Raman spectroscopy analysis 91

**Chap V.   Effets de l'hydratation sur les propriétés mécaniques du *stratum
corneum* 117**

Article 2 : The relationship between water loss, mechanical stress, and molecular
structure of human *stratum corneum ex vivo* 119

**Chap VI.   Détermination de descripteurs du stress mécanique sur le *stratum
Corneum* : spectroscopie Raman et PLS 137**

Article 3 : Raman spectroscopy : a tool for biomechanical charactérisation of *stratum
corneum*.. 140

**Chap VII.   Effets des molécules hydratantes sur la structure du *stratum corneum* :
Spectroscopie vibrationnelles et CLS 163**

VII.1.   Préambule.. 164

Article 4 : Vibrational spectroscopy coupled to classical least square analysis, a new
approach for determination of skin moisturizing agents' mechanisms 167

Chap VIII. Caractérisation *in vivo* de la peau par spectroscopie Raman 191

Article 5 : Raman spectroscopy: in vivo Quick Response "QR" code of skin physiological
status 194

CONCLUSION GENERALE ET PERSPECTIVES...222

REFERENCES 225

11

LISTE DES ABBREVIATIONS

ACP	Analyse en Composantes Principales
CCD	Charged Coupled Device
CLS	Moindres carrées classiques
DSC	Calorimétrie différentielle à balayage
EI	Impédance Electrique
IRM	Imagerie par Résonance Magnétique
LB	Corps lamellaire
NMF	Facteurs naturels d'hydratation
OCT	Tomographie par Cohérence Optique
PCA	Acide Carboxylique Pyrrolidone
PIE	Perte Insensible en Eau
PLS	Regression par moindres carrés partiels
PLS-DA	Analyse discriminante par moindres carrés partiels
SC	*stratum corneum*
SB	*stratum basale*
SG	*stratum granulosum*
SS	*stratum spinosum*
RMN	Résonnance Magnétique Nucléaire
UCA	Acide Urocanique

LISTE DES FIGURES

Figure 1: Structure générale de la peau humaine 22

Figure 2: Représentation schématique du *stratum corneum* et de sa structure lipidique. (A) Modèle «briques et mortier» de l'organisation du *stratum corneum*. (B) Photomicrographie de cornéocytes isolés. (C) Domaines et modèle mosaïque de la barrière lipidique. (D) Lipides en phase gélifiée avec des chaînes hydrocarbonées qui adoptent une conformation tous-trans et de domaines liquides ayant des chaînes hydrocarbonées qui adoptent des conformations gauches (microstructure de la barrière lipidique). (E) Organisation latérale de la chaîne hydrocarbonée [9]. 23

Figure 3: Modèle de la bicouche lipidique du stratum corneum : (a) CERamides pure, (b) mélange équimolaire ceramides-acides gras (1:0:1), (c) mélange équimolaire ceramides-cholestérol (1:1:0) et (d) mélange des ceramides- cholestérol et acides gras (2:2:1) à 340 K. Les ceramides, le cholestérol, et les acides gras sont représentés par les lignes épaisses et colorées en bleu, orange et vert, respectivement. Les molécules d'eau sont illustrées par les lignes fines [22]. 25

Figure 4: Principe de mesure d'un dispositif à chambre ouverte: les instruments de mesure à chambre ouverte sont basées sur la loi de la diffusion de Fick, indiquant la quantité transportée par unité de surface définie et par période de temps. L'air contenant de l'eau se déplacé le long des deux capteurs à l'intérieur de la tête de mesure. En utilisant les données obtenues de thermo-et hydro-capteurs, et après le traitement des informations par un microprocesseur intégré, on obtient une valeur numérique de la perte d'eau trans-épidermique, généralement représenté en $g/m^2/h$ [55] 33

Figure 5: Dispositif pour l'évaluation de l'hydratation utilisant la mesure d'impédance: La mesure est basée sur la différence des constantes diélectriques de l'eau et d'autres substances dans le champ de mesure électrique. La faible fréquence de fonctionnement (40 à 75 Hz) de la sonde est sensible aux changements relatifs de la constante diélectrique du SC [55]. 35

Figure 6: Courbes typiques de forces d'allongement pour différentes niveau d'humidité relative montrant les différentes phases. obtenu à partir de Wildnauer et al.[79]. 39

Figure 7: Modes de vibration d'un groupement plan CH_2 : les schémas du dessus représentent les vibrations d'élongation symétrique (gauche) et asymétrique (droite). Les schémas du milieu représentent les vibrations de déformation, dans le plan, de cisaillement (gauche) et de balancement (droite). Enfin, les schémas du bas représentent les vibrations de déformation, hors du plan, de torsion (gauche) et de balancement (droite). 45

Figure 8: Diagramme de Jablonsky : evolution de l'énergie de vibration résultant de l'interaction de la lumière avec la matière. Adapté de [113] 47

Figure 9: Spectre Raman du silicium (longueur d'onde d'excitation 632,8 nm) [114]..... 50

Figure 10: Schemas du microspectromètre Raman LabRam 51

Figure 11: Rendement quantique des détecteurs (CCD et InGaAs) en fonction de la longueur d'onde. Étendue en nm de la gamme spectrale (100-3000 cm^{-1}) en fonction de l'excitatrice [114]. 53

Figure 12: Système micro-sonde Raman: (a) source laser, (b) Charged Coupled Device (CCD), (c) micro hr, (d1 et d2) shutter laser et shutter spectrometre respectivement, (f) fibre permettant une confocalite de 5 μm et (e) micro-sonde confocale 54

Figure 13: Schéma de principe de l'analyse par spectroscopie d'absorption infrarouge ; Appareil de type Spectrum two (Perkin Elmer) 59

Figure 14: Dispositif ATR équipé d'un cristal diamant 61

Figure 15: Dispositif d'essai pour la technique de la courbure du substrat illustrant le *stratum corneum* collé sur un substrat de verre : un laser a balayage equipe d'un détecteur mesure l'angle de déflexion (α) par rapport a la position (y) sur le substrat ; la courbure moyenne est calculée a partir d'une régression linéaire de l'angle de déviation [3] 63

Figure 16: Relative Humidity Controler 01 (RHC_01) 74

Figure 17: Controleur d'Humidité relative 02 (RHC_02) 75

Figure 18: Spectres Raman du *stratum corneum*. 1) spectre obtenu avec une source laser 633 nm, un réseau de 1800 traits/mm. 2) spectre obtenu avec une source laser de 660nm, un réseau de 950 traits/mm 82

Figure 19: Molécules d'eau sous forme tétraédrique 86

LISTE DES TABLEAUX

Tableau 1: Caractéristiques de principaux lasers utilisées en spectroscopie Raman 52

Tableau 2 : Pression de la vapeur d'eau en fonction de la température 69

Tableau 3: Influence des sels sur les pressions de vapeur et de HR 70

Tableau 4: Humidités relatives de solutions saturées de sels 74

Tableau 5: Réglage de HR sur le système RHC_02 77

Tableau 6: Paramétres d'acquisition spectrale, LabRam HR 80

Tableau 7: Paramètres d'acquisition spectrale, LabRam 81

Tableau 8: Etude de la variation du signal en fonction du temps et du point analysé.
Analyse ANOVA 84

INTRODUCTION GENERALE

La peau représente le premier contact entre le corps humain et son environnement. Son aspect physique est aussi important sur le plan esthétique que fonctionnel. L'une des principales fonctions de la peau est de fournir une barrière compétente contre la perte d'eau et de maintenir un niveau d'hydration requis pour une fonction physiologique normale équilibrée. Cette régulation est cruciale car elle ne concerne pas seulement l'hydratation mais présente aussi un impact sur les propriétés mécaniques et l'aspect de la peau.

Bien que la répartition de l'eau dans la peau et le gradient d'hydratation soient étroitement régulés, une partie de l'eau du corps est continuellement perdue à partir des couches les plus externes de la peau vers l'atmosphère. Ce contrôle est assuré par la préservation de l'intégrité de la barrière cutanée représentée essentiellement par le *stratum corneum* (SC). L'intégrité de la barrière est souvent compromise par l'action de facteurs extrinsèques comme les perturbations environnementales, les agressions chimiques, mécaniques, ou par l'exposition au soleil. Cependant, des mécanismes intrinsèques fonctionnels dans la peau permettent d'assurer qu'un niveau suffisant d'hydratation soit maintenu. Néanmoins, des facteurs physiologiques tels que l'âge et le statut hormonal, ou pathologiques de la peau peuvent compromettre, dans certains cas, le rétablissement de niveaux d'hydratation cutanée adéquats. Ainsi, un grand pourcentage de la population pourrait être potentiellement affecté par un état de "sécheresse cutanée". En outre, de nombreuses maladies de la peau sont accompagnées par une perturbation de la fonction barrière et par un assèchement de la peau. Il est donc important d'identifier avec précision les mécanismes moléculaires impliqués dans ce phénomène afin de pouvoir lutter efficacement contre les divers troubles se manifestant par une peau sèche.

En cas de sécheresse cutanée à la suite d'un affaiblissement des facultés de la réhydratation, il est nécessaire d'aider la peau à retrouver son état d'hydratation optimal notamment par des applications topiques de « substances hydratantes ». Toutefois, il existe peu de modèles prédictifs précliniques pertinents en mesure de répondre aux normes élevées requises pour tester l'efficacité de ces produits. Ainsi, une méthode d'analyse *ex vivo* capable de mimer l'état de sécheresse de la peau *in vivo* en réponse à un stress environnemental est nécessaire.

Le SC constitue la dernière zone frontière entre l'organisme et son environnement. D'une part, elle représente la région dans laquelle les effets dus aux variations environnementales ou aux traitements topiques sont détectés, et d'autre part la zone dans laquelle le résultat symptomatique de la barrière pourrait être observé. Sur un SC isolé *ex vivo*, ce résultat ne reflète pas l'action des ingrédients sur la régulation moléculaire des éléments constituant les couches vivantes de l'épiderme, mais est plutôt la preuve de leur action topique locale. La proposition de solutions analytiques capables de montrer cette action locale *ex vivo* par rapport à des données *in vivo* permettrait la discrimination de molécules hydratantes adaptées à chaque indication. Parmi les approches d'intérêt utilisé dans l'étude du SC, les techniques d'analyse physico-chimiques semblent adaptables à cette problématique. En particulier, les techniques d'analyses spectroscopiques et biomécaniques présentent un intérêt potentiel pour la caractérisation de l'état d'hydratation du SC. La déshydratation du SC est associée à la réduction du volume et hypothétiquement à divers mécanismes non encore élucidés tels que, la modification de sa structure [1, 2]. Ces phénomènes conduisent à la modification de ses propriétés biophysiques comme l'épaisseur, l'élasticité [3] et la tension qui à leur tour influent sur son intégrité.

En plus de leur capacité à mesurer l'hydratation, les techniques spectroscopiques vibrationnelles sont non-destructives et non-invasives et permettent l'accès à l'information moléculaire et structurale très spécifique concernant l'échantillon étudié sans avoir besoin d'aucune préparation ou marquage spécial de l'échantillon. Ainsi, parmi toutes les techniques non invasives de mesure développées récemment, la spectroscopie Raman confocale se démarque des autres parce qu'elle permet de récupérer des données tridimensionnelles *in vivo*, en ajoutant le paramètre profondeur concernant les mesures des profils de concentration de l'eau, des NMF et de lipides [2, 4, 5].

D'autre part, les conséquences du niveau d'hydratation cutanée sur l'homéostasie épidermique peuvent être évaluées *in vivo* à la surface de la peau du volontaire par échantillonnage non-invasif. A partir d'un prélèvement des substances biologiques du SC, il est possible de mesurer l'activité de certaines enzymes et d'effectuer des analyses biochimiques. En effet, les techniques d'analyse chromatographiques couplée à la spectrométrie de masse permettent la détection des NMFs et le dosage des différents lipides impliqués dans le maintien de la barrière. Grâce à toutes ces techniques, il parait possible d'identifier ces descripteurs moléculaires et mécaniques liés à l'hydratation de la peau.

Dans ce travail de thèse, nous avons donc développé une méthodologie basée sur l'analyse de l'hydratation du SC en utilisant les technologies analytiques qui sont transférables ou déjà utilisé *in vivo* comme la micro-spectroscopie Raman. Le manuscrit est composé de différentes parties décrivant l'apport de ce travail dans l'amélioration de la connaissance du comportement du SC placé dans des conditions de sécheresse. Nous décrirons d'abord l'état de l'art sur la peau sèche ainsi que sur les différentes techniques d'analyse de la sécheresse cutanée et des propriétés mécaniques du SC. On y évoque le principe ainsi que la théorie de fonctionnement de chacune de ces techniques pour finir par leurs applications dans l'analyse du SC. Les performances, les avantages ainsi que les limites de chaque technique sont discutées.

Le cœur de cette thèse consiste à caractériser au niveau moléculaire l'hydratation du SC. Dans cette partie, sont mises en évidence les modifications structurales du SC engendrées par un changement d'HR ambiante. Les changements de l'ultra-structure de l'eau, des lipides et des protéines du SC y sont décrits. Puis, une contribution est apportée sur l'effet de l'hydratation sur les propriétés mécaniques du SC tout en mettant en évidence les mécanismes fonctionnels impliqués dans l'apparition du stress lié à la déshydratation. Les descripteurs spectroscopiques témoin de l'extension mécanique du SC sont également identifiés à partir d'une approche chimiométrique appliquant la régression par moindres carrés partiels (Partial Least Square : PLS) sur les données acquises en spectroscopie Raman.

A côté de ces investigations fondamentales, les différents mécanismes d'action des molécules hydratantes ont été explorés par les spectroscopies vibrationnelles couplées à la régression classique par moindres carrées (Classical Least Square : CLS) afin d'isoler les variations uniquement liées au SC. Cette approche permet l'attribution et la quantification de l'activité de réhydratation spécifique à une molécule donnée.

Enfin, sur base des études *in vivo*, nous proposons une nouvelle approche d'analyse multiparamétrique couplant la microspectroscopie Raman au traitement PLS comme méthode permettant d'avoir des informations sur la composition lipidique du SC, l'hydratation et la structure protéique à différentes profondeurs ainsi que les mesures biométriques comme le pH et la perte insensible en eau.

Etat de l'art

CHAP I. *STRATUM CORNEUM*:
STRUCTURE, FONCTIONS ET
PROPRIETES

I.1. STRUCTURE DU *STRATUM CORNEUM*

Le SC fait partie intégrante de la peau qui elle-même est constituée de trois couches principales : l'hypoderme, le derme et l'épiderme (Figure 1). Ce dernier est la couche la plus externe et la plus fine de la peau, très résistante, kératinisée et non vascularisée.

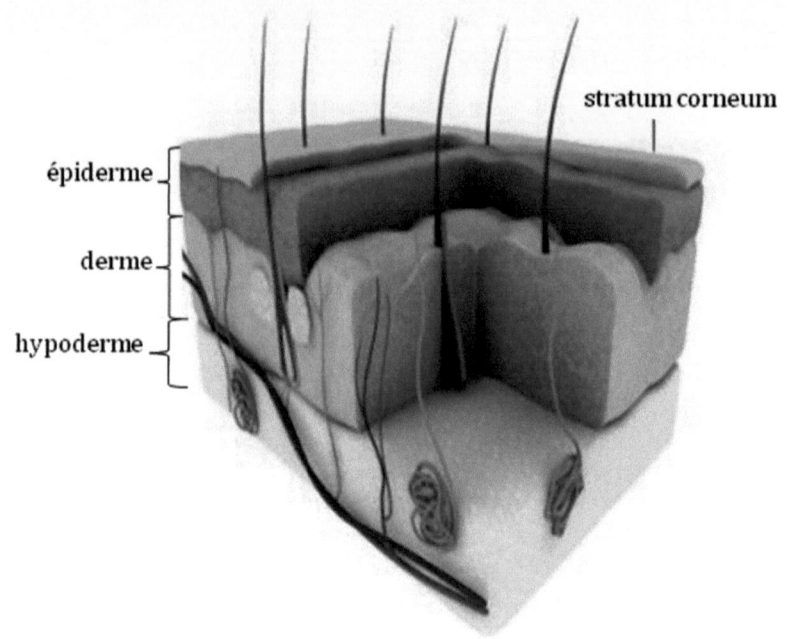

FIGURE 1: STRUCTURE GENERALE DE LA PEAU HUMAINE

Le SC représente à son tour la couche la plus superficielle de l'épiderme. Il a une structure globale relativement simple avec des cellules appelées cornéocytes reliées par des desmosomes. L'ensemble est imbibé dans une matrice de lipides qui rempli l'espace intercellulaire [6, 7]. Cela a incité sa comparaison à une structure en "brique et mortier" (Figure 2A) originairement décrite par Michaels et al. [8]; où les cornéocytes forment les «briques» et les lipides extracellulaires représentent le «mortier». Ces couches de cellules avec leur connections représentent entre 75 et 80% du volume de SC. Les 25-20% restants correspondent à la matrice lipidique. L'épaisseur entière du SC est évaluée entre 15-20 µm.

Au niveau ultrastructural, le SC présente plutôt une incroyable complexité structurelle et fonctionnelle à la fois dans les briques (Figure 2B) et dans les composants du mortier (Figure 2C). La Figure 2 illustre les principales composantes structurelles de ce système composite :

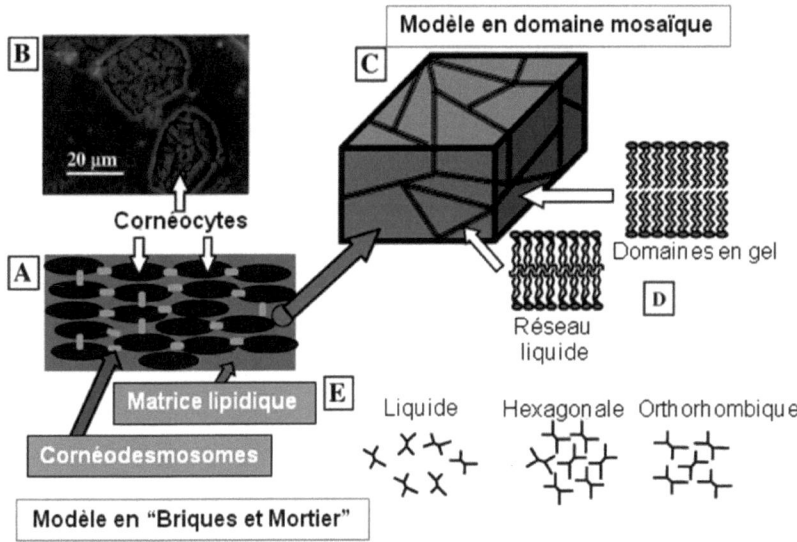

FIGURE 2: REPRESENTATION SCHEMATIQUE DU *STRATUM CORNEUM* ET DE SA STRUCTURE LIPIDIQUE. (A) MODELE «BRIQUES ET MORTIER» DE L'ORGANISATION DU *STRATUM CORNEUM*. (B) PHOTOMICROGRAPHIE DE CORNEOCYTES ISOLES. (C) DOMAINES ET MODELE MOSAÏQUE DE LA BARRIERE LIPIDIQUE. (D) LIPIDES EN PHASE GELIFIEE AVEC DES CHAINES HYDROCARBONEES QUI ADOPTENT UNE CONFORMATION TOUS-TRANS ET DE DOMAINES LIQUIDES AYANT DES CHAINES HYDROCARBONEES QUI ADOPTENT DES CONFORMATIONS GAUCHES (MICROSTRUCTURE DE LA BARRIERE LIPIDIQUE). (E) ORGANISATION LATERALE DE LA CHAINE HYDROCARBONEE [9].

> **Les cornéocytes** (Figure 2B): ce sont des cellules plates anucléées qui sont sous la forme hexagonale ou pentagonale empilées jusqu'à 15-20 couches en fonction de la localisation anatomique dans le corps. Leur largeur est comprise entre 20-30 μm et leur épaisseur est de 0,5-1,0 μm [10]. En plus des filaments de kératine les corneocytes contiennent la filaggrine, dont la protéolyse fourni les acides aminés constitutifs ainsi que les sels dérivés d'acides aminés [11] le mélange ainsi obtenu, appelé facteurs naturels d'hydratation (NMF), est très soluble dans l'eau et peut représenter autour de 10% du poids sec des corneocytes [12]. Ces substances sont enveloppées dans une membrane protéique cornée semi-rigide.

Ces cellules cornées constituent un réservoir d'eau du SC et contribuent aussi à la formation de la barrière physique à la pénétration.

➤ **Les cornéodesmosomes** fonctionnent comme des "rivets" pour maintenir les cornéocytes ensemble. Ces réseaux de protéines sont spécialisés dans des liens inter-corneocytaires. Les desmosomes sont programmés pour passer par un processus de dégradation progressive de manière à permettre la desquamation ordonnée ultrapériphériques des corneocytes usés [13].

➤ **Les lipides** du mortier remplissent les espaces entre les corneocytes et sont constitués d'un mélange très complexe de céramides, de cholestérol et d'acides gras libres dans des proportions très variables entre individus et en fonction du site anatomique analysé [14] ; l'organisation supramoléculaire de ces lipides constitue l'élément majeur de la propriété barrière cutanée.

Le cholestérol est essentiel pour l'organisation du mélange des différentes espèces lipidiques, il assure également la fluidité et la souplesse au SC [15].

Les Acides gras libres se composent principalement de longues chaînes d'acides gras saturés ayant plus de 20 atomes de carbone. L'acide oléique C18 :1(6%) et de l'acide linoléique C18 :2(2%) sont les seuls acides gras insaturés détectés non liés aux membranes des cellules corneocytaires du SC.

Les céramides ont des longueurs de chaîne entre 20 et 36 atomes de carbone, la majorité ayant 26-28 atomes de carbones (exception pour céramide AS qui a une chaîne moyenne de C15-C18) [16]. Ces sphingolipides sont composées d'acides gras qui peuvent être de trois types (non-hydroxylés (N), α-hydroxylés (A) et ω-hydroxylés (EO)) et une base sphingoside variable de types (sphingosine (S), phytosphingosine (P) et 6-hydroxysphingosine (H), dihydrosphingosine (dS)) [17]. Des études récentes indiquent l'existence de 12 classes de céramides détectés dans la matrice lipidique du SC humain [18].

Ces chaînes traversent la bicouche lipidique et se lient au niveau de la partie linoléique, par des interactions hydrophobes de type de Van der Waals, à des molécules de cholestérol au sein d'une couche lipidique fluide séparant 2 bicouches superposées (Figure 3). Il se forme ainsi une véritable "matrice" lipidique qui est responsable de l'effet barrière vis-à-vis des molécules, tout en laissant la possibilité de passage d'un certain flux d'eau (Figure 2D et 3); cette "matrice" est consolidée par les interactions latérales des très longues chaînes qui constituent de véritables "rivets moléculaires" accroissant la stabilité et la cohésion de l'ensemble [19-21].

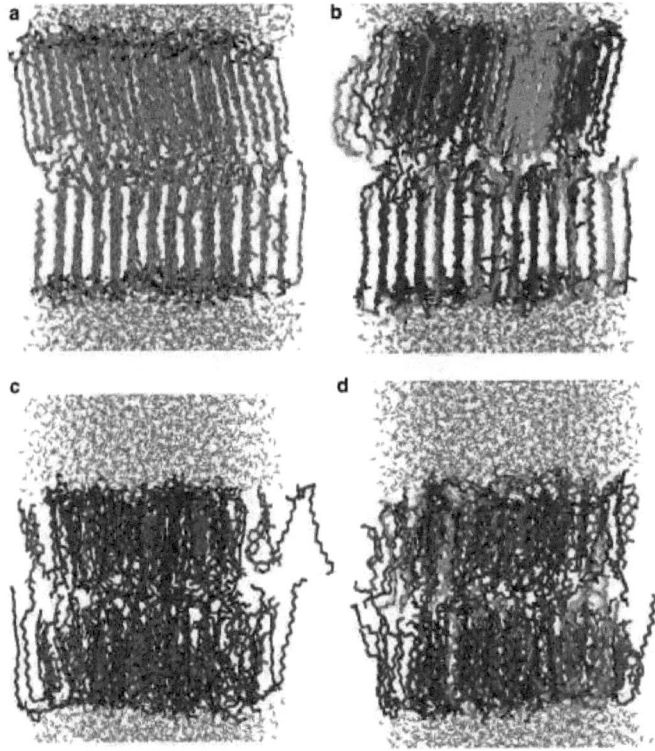

FIGURE 3: MODELE DE LA BICOUCHE LIPIDIQUE DU STRATUM CORNEUM : (A) CERAMIDES PURE, (B) MELANGE EQUIMOLAIRE CERAMIDES-ACIDES GRAS (1:0:1), (C) MELANGE EQUIMOLAIRE CERAMIDES-CHOLESTEROL (1:1:0) ET (D) MELANGE DES CERAMIDES- CHOLESTEROL ET ACIDES GRAS (2:2:1) A 340 K. LES CERAMIDES, LE CHOLESTEROL, ET LES ACIDES GRAS SONT REPRESENTES PAR LES LIGNES EPAISSES ET COLOREES EN BLEU, ORANGE ET VERT, RESPECTIVEMENT. LES MOLECULES D'EAU SONT ILLUSTREES PAR LES LIGNES FINES [22].

Les études en diffraction des rayons X ont révélé des informations concernant l'organisation des lipides dans le SC. Ces couches présente deux phases lamellaires avec des distances de répétition de 6 à 13mm, ce dernier étant la majorité [23]. En outre, les lipides possèdent trois sortes d'assemblages latéraux (Figure 2E) : désordonné, hexagonale ou orthorhombique, avec une domination de ce dernier. Les céramides sont également responsables de la formation de l'enveloppe lipidique liée par covalence aux cornéocytes [24, 25].

➢ **Un ensemble d'enzymes lipolytiques et protéolytiques** impliquées dans la libération des lipides et dans la dégradation des desmosomes respectivement. Elles contribuent aux activités biochimiques se produisant dans le SC, qui était autrefois considéré comme inerte et mort. Les enzymes protéolytiques sécrétées par les corps lamellaires (Lamellar Body : LB) conduisent à une dissolution progressive des cornéodesmosomes permettant ainsi la desquamation ordonnée des cornéocytes. Plus récemment, des peptides anti-microbiens comme les bêta défensines et cathélicidines [26] ont été co-localisés dans les corps lamellaires et trouvés au sein de la matrice extracellulaire du SC, reliant ainsi la formation de la barrière au molécules à la barrière anti-microbienne [27]. Cette immunité innée épithéliale, partie prénante de la capacité de la peau à se défendre vis-à-vis de l'extérieur, est cruciale pour la survie. De plus, l'application de simples formulations émollientes a montré une modification de l'activité enzymatique desquamatoire et inflammatoire dans le SC [28].

➢ **Les petites quantités de substances sécrétées** provenant des corps lamellaires épidermiques à l'interface du SC et de la couche granuleuse. Ce sont les composants lipidiques (phospholipides et glucosylceramides) précurseurs des différentes classes lipidiques du SC mais aussi des enzymes et peptides antimicrobiens.

Tous ces éléments sont essentiels à la barrière du SC ; interférer avec ou modifier les propriétés fonctionnelles de l'un de ces composants peut affaiblir la barrière.

Dans les années 1970, le SC était considéré comme une matière inerte composée de cellules totalement différentiées, aplaties et énucléés remplies de kératine et délimitées par une enveloppe cornifiée. Les études qui ont suivi ont démontré que la description du SC comme une membrane d'occlusion passive de l'épiderme était erronée [29]. Le rôle de la protéine « filaggrine » dans les cornéocytes ainsi que sa digestion pour former les NMF est d'une importance capitale dans le bon fonctionnement du SC [30, 31].

I.2. HYDRATATION CUTANEE ET FONCTION BARRIERE

Le derme est continuellement alimenté en eau par la voie de la circulation sanguine. Dans le derme, une partie de l'eau est présente sous forme de gel, fixé à de nombreuses macromolécules hydrophiles. En dépit de la grande capacité de ces molécules à capter l'eau, une partie reste mobile et diffuse vers l'épiderme qui sert de première barrière de protection à l'environnement extérieur. En plus de l'eau perdue par la transpiration, une partie d'eau est perdue par évaporation. Cette Perte Insensible en Eau (PIE) est d'environ $5g/m^2.h$ dans des conditions physiologiques normales. Ces mouvements d'eau contribuent au bon fonctionnement du SC par approvisionnement continu en eau. Entre le derme fortement hydraté et l'épiderme, un gradient décroissant existe et devient plus évident entre les couches épidermiques les plus profondes et le SC. En particulier, alors que toutes les couches vivantes de l'épiderme (*stratum basale*, SB, *stratum spinosum*, SS, et *stratum granulosum*, SG) présentent un niveau d'hydratation élevé (de l'ordre de 70%, à poids égal), une chute brutale de l'hydratation suit la transition entre SG et SC. A ce niveau, le taux d'hydratation chute à 30% puis à 15% environ dans les couches les plus superficielles.

I.2.1 REGULATION DE L'HYDRATATION DU *STRATUM CORNEUM*

Dans le SC caractérisé par des cellules différenciées très spécifiques, il ya deux éléments principaux de régulation de la perméabilité à l'eau: a) l'adsorption de l'eau par des substances hydrophiles (NMF) qui constitue une barrière "statique", et b) la limitation de la diffusion de l'eau à travers le SC grâce à la présence d'une matrice lipidique inter-cornéocytaire qui représente une barrière hydrophobe plus dynamique. Comme mentionné dans la section précédente, cette dernière limitation et sa structuration particulière jouent un rôle important et indispensable dans la régulation des flux d'eau.

I.2.1.1 ABSORPTION D'EAU DANS LE *STRATUM CORNEUM*

Etant hygroscopiques, les NMF, glycérol [32] et l'acide hyaluronique fixent l'eau dans le SC et absorbent l'eau atmosphérique fournissant une hydratation suffisante pour aider à garder la peau souple et facilite diverses réactions enzymatiques [33].

27

Les NMF sont contenus dans les cornéocytes. Ils sont essentiellement composés d'acides aminés libres (40%) et leurs dérivés tels que l'Acide Carboxylique Pyrrolidone (PCA) et de l'acide urocanique (UCA). Les sels d'acide lactique (12%), importants pour le maintien de l'acidité à la surface du SC, l'urée (8%) et différents sels minéraux sont également présents. La capture de l'eau est due en partie à une forte pression osmotique qui augmente avec la différenciation des cornéocytes, associé à une augmentation de la concentration d'acides aminés, de métabolites de faibles masses moléculaires comme les lactates et autres à l'intérieur des corneocytes. Cette différence de pression intra-et extra-cellulaire conduit à un transfert d'eau vers l'intérieur des cornéocytes. La filaggrine, nécessaire à la synthèse et la libération des acides aminés des NMF, est une protéine essentielle pour l'hydratation cutanée. Elle est produite dans les kératinocytes du SG sous la forme de la profilaggrine précurseur. Au niveau SG-SC, les sous-unités de filaggrine subissent une protéolyse qui va générer les acides aminés constituant les NMF [34].

I.2.1.2   BARRIERE A LA DIFFUSION DE L'EAU

Le rôle de la matrice lipidique est essentiel dans la régulation de la perméabilité à l'eau. Il représente une barrière hydrophobe à la désorption de molécules d'eau à travers les espaces inter-cornéocytaires. Les lipides présents dans la matrice inter-cornéocytaire: les céramides, les acides gras libres et le cholestérol sont synthétisés à partir de précurseurs lipidiques libérés à partir des LB, à la jonction SG-SC. Ainsi, dans des conditions physiologiques normales, les céramides, les acides gras et le cholestérol forment un réseau supra-moléculaire organisé de façon optimale afin d'assurer un flux d'eau trans-épidermique modéré. Parmi les lipides impliqués dans la formation de la matrice lipidique extracellulaire et par conséquent, à l'écoulement d'eau à travers le SC, les céramides jouent un rôle majeur. La composition qualitative et quantitative en céramides au sein du SC est très spécifique permettant un contrôle de l'organisation fonctionnelle de la matrice lipidique.

D'une manière générale, les modifications quantitatives et/ou qualitatives de n'importe quel lipide de ces trois classes se traduit par des anomalies de la barrière cutanée qui se caractérisent par une augmentation de la PIE ainsi que des modifications organisationnelles supramoléculaires de la matrice lipidique extracellulaire [35].

I.2.1.3   SYSTEME DE CONTROLE ET DE REPONSE ADAPTATIVE

Tous les paramètres décrits ci-dessus correspondent aux conditions physiologiques non-perturbées assurées par une homéostasie finement régulée. Mais qu'advient-il quand le niveau d'hydratation est perturbé? Deux situations peuvent se présenter:

(a) l'organisme est encore capable de produire une réponse adaptative dans le but de rétablir l'équilibre physiologique en eau, la fonction barrière et de préserver les propriétés fonctionnelles et de réactivité dynamiques du SC. Ce processus d'adaptation est en mesure de répondre aux variations de l'environnement (température, humidité, exposition aux ultraviolets, la pollution, les détergents, etc.);

(b) dans une situation physiopathologique accompagnée par un défaut dans l'adsorption de l'eau, et de la fonction barrière (l'âge, le statut hormonal, maladies de la peau). Dans ce cas la capacité d'initier une réponse adaptative à une perturbation est réduite, voire inexistante, et doit donc être compensée.

Une série de réponses homéostatiques sont activées afin de rétablir une barrière efficace et la capacité hygroscopique du SC. Tout d'abord une sécrétion rapide des lipides dans les LB a lieu afin de reconstituer le "ciment lipidique". Lorsqu'un individu est confronté à des variations environnementales d'HR ou de température suite à des changements climatiques, la peau réagit par un renforcement de la fonction de barrière. Par exemple, dans des environnements secs, une augmentation de la synthèse des céramides et des composantes des NMF, permet de limiter la diffusion en eau et d'améliorer la capacité à fixer l'eau dans le SC. Les études sur la distribution de l'eau dans le SC par spectroscopie Raman *in vivo* ont démontré que selon la saison, les variations se produisent principalement dans les couches les plus superficielles [4].

I.2.2   CARACTERISTIQUES D'UNE PEAU SECHE

Dans certains cas, la réponse adaptative de l'organisme sera insuffisante ou déréglée. Cela conduira à l'assèchement de la peau. La peau sèche ou xérose est un état physiologique cutané caractérisé par l'accumulation de cornéocytes à la surface de la peau (une peau squameuse), donnant une texture rugueuse [36]. L'apparition de la peau sèche dépend de nombreux facteurs extrinsèques tels que le climat, l'environnement et l'exposition à des savons, à des détergents, à des produits chimiques ou à des médicaments. Il existe également une variété de facteurs intrinsèques qui peuvent contribuer à cet état, tel que la génétique, le déséquilibre hormonal, les maladies, et le vieillissement [21]. Cliniquement, les signes et les symptômes de sécheresse cutanée peuvent se manifester par une diminution concomitante de la flexibilité, une sensation de raideur au toucher, l'existence de prurit, une dérégulation de la desquamation, des irritations, des érythèmes, des douleurs et des picotements [36]. Dans certains cas ces symptômes sont accompagnés par des fissures ou des démangeaisons. Dans presque tous cas, la peau xérotique est rude, elle manque de souplesse et est terne en apparence. Par ailleurs, une modification de la teneur en céramides du SC a été montré [37] induisant des changements dans les propriétés physico-chimiques des lipides lamellaires avec une désorganisation de la phase lamellaire et une organisation orthorhombique réduite [38]. On observe également une diminution de certains composants des NMF. Cela se traduit par une capacité réduite à retenir l'eau, une augmentation de la PIE et de la perméabilité du SC. La peau sèche se produit principalement sur les jambes mais peut affecter la surface entière de la peau.

Au cours des dernières décennies, notre compréhension des mécanismes impliqués dans la sécheresse cutanée a progressé énormément. Malgré ces progrès, la peau sèche reste le plus commun des troubles de la peau humaine [39]. Une meilleure compréhension des aspects moléculaires de la peau sèche est donc d'une grande importance en dermo-cosmétique non seulement pour un diagnostic précis mais aussi pour développer de nouveaux traitements et mesurer les effets du traitement.

I.2.3   GESTION DE LA PEAU SECHE: PRODUITS HYDRATANTS

La perturbation du processus physiologique permettant le maintien de la barrière cutanée provoque une série de réactions en cascade impliquant entre autres : la perte de NMF, la réduction des activités enzymatiques dont celles responsables de la libération des lipides du SC, puis la dérégulation d'un certain nombre de processus cellulaires (inflammation, prolifération et différenciation épidermique) [37], et enfin l'apparition des signes phénotypiques mentionnés dans le paragraphe précédant. Ainsi, le cycle de peau sèche est auto-entretenu [40, 41]. Il est donc nécessaire de gérer cette peau sèche pour éviter que cycle progresse encore plus.

L'objectif est donc d'améliorer l'hydratation cutanée en compensant ou en stimulant la réparation endogène de la barrière épidermique et sa capacité hygroscopique, en utilisant des substances lipophiles ou hydrophiles ou émulsions aqueuses enrichies avec des composants hygroscopiques.

Deux types principaux d'agents hydratants existent classiquement. Les agents occlusifs s'opposent à la déshydratation en formant un film de lipides à la surface de la peau qui limite la PIE. Ce sont tous des lipides avec des hydrocarbures. Les plus couramment utilisés sont : vaseline, l'huile de paraffine, le perhydrosqualène, les huiles de silicone, les huiles animales et végétales, les alcools gras, les cires (beurre de karité, la lanoline, etc...). Ces produits ne sont pas utilisés seuls mais sous la forme d'émulsions eau-dans-huile. Les agents humectants sont très hygroscopiques (ont une forte affinité pour l'eau): ils comprennent les NMF ou des complexes constitués de ces éléments les plus actifs, notamment le lactate de sodium, sels de sodium de l'acide pyrrolidone carboxylique et l'urée. Les polyols, tels que l'éthylène glycol, le glycérol ou le propylène glycol sont également utilisés en tant qu'agents humectants. A ces deux types majeurs s'ajoutent les modulateurs de la barrière cutanée mal décrits dans la littérature. Ces molécules changent la conformation et l'organisation des lipides du SC en leur conférant une meilleure compacité. Les additifs présents dans les produits hydratants sont multiples mais leur impact sur la peau est discutable quand il s'agit d'une application topique. Certains outils sont disponibles pour évaluer l'efficacité des formulations hydratantes.

I.2.4   METHODES D'EVALUATION OBJECTIVE DE LA PEAU SECHE

Dans les dernières décénies, il y a eu une explosion dans le développement d'outils de recherche disponibles pour l'exploration de nombreux paramètres relatifs à l'état, à la fonction et à la réactivité de la peau aux perturbations [2]. Généralement l'évaluation de la peau sèche s'effectue avec les méthodes électriques comme l'Impédance Electrique (EI), pour la mesure d'hydratation du SC et la PIE [14, 42-45]. Des techniques relativement nouvelles, avec un énorme potentiel pour la description des processus anormaux sont en plein essor. Parmi ces diverses méthodes, on compte: la spectroscopie Raman [4, 46-50], la microscopie confocale [51], la tomographie par cohérence optique (OCT) [52] et l'imagerie par résonance magnétique (IRM). Les 3 premières techniques sont discutées dans cette section en fonction de leur utilité à évaluer objectivement l'hydratation de la peau et identifier les anomalies épidermiques.

I.2.4.1   LA PERTE INSENSIBLE EN EAU

La mesure de la PIE est un outil non-invasif important fréquemment utilisé pour évaluer les changements dans la fonction barrière du SC. La mesure de la PIE révèle la diffusion passive de l'eau et celle relative à la transpiration. La plupart des études ont montré que l'augmentation de la PIE est souvent liée à une fonction barrière altérée [39, 53]. A contrario, une diminution de la PIE reflète généralement la restauration de cette barrière [43].

Les mesures de PIE sont basées sur l'estimation du gradient de la vapeur d'eau dans une chambre de mesure [54]. La concentration de l'eau dans la peau et le flux peut être liée par la loi de Fick:

$$J = -D \bullet \frac{dc}{dx} \qquad\qquad \text{ÉQUATION 1}$$

Où

J = Flux de diffusion (g/m.h)

D = coefficient de diffusion

c = concentration de l'eau

x = Distance de la surface de la peau au point de mesure

dc / dx = gradient de concentration d'eau

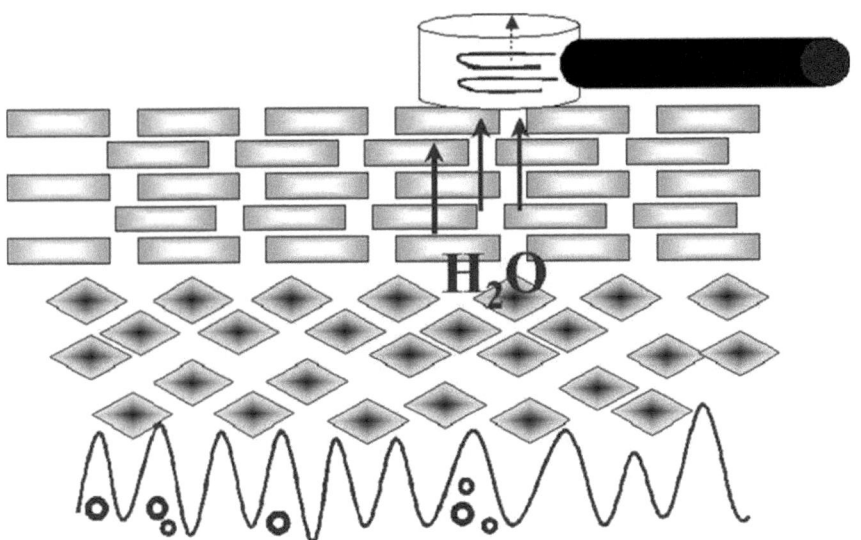

FIGURE 4: PRINCIPE DE MESURE D'UN DISPOSITIF A CHAMBRE OUVERTE: LES INSTRUMENTS DE MESURE A CHAMBRE OUVERTE SONT BASEES SUR LA LOI DE LA DIFFUSION DE FICK, INDIQUANT LA QUANTITE TRANSPORTEE PAR UNITE DE SURFACE DEFINIE ET PAR PERIODE DE TEMPS. L'AIR CONTENANT DE L'EAU SE DEPLACE LE LONG DES DEUX CAPTEURS A L'INTERIEUR DE LA TETE DE MESURE. EN UTILISANT LES DONNEES OBTENUES DE THERMO-ET HYDRO-CAPTEURS, ET APRES LE TRAITEMENT DES INFORMATIONS PAR UN MICROPROCESSEUR INTEGRE, ON OBTIENT UNE VALEUR NUMERIQUE DE LA PERTE D'EAU TRANS-EPIDERMIQUE, GENERALEMENT REPRESENTE EN $G/M^2/H$ [55].

De nombreuses variables doivent être prises en considération lors de l'interprétation de la PIE, y compris les variations anatomiques, l'âge, le sexe, la race, l'activité de la glande sudoripare, le rythme circadien, l'HR, la température de la sonde de mesure [43, 45, 56]. La mesure de la PIE est une méthode largement utilisée, mais la variabilité des conditions opératoires qui engendrent des résultats contradictoires ont empêché la comparaison entre les différentes études. Plusieurs instruments ont été conçus pour mesurer la PIE. Les techniques les plus utilisées peuvent être divisées en deux principales catégories : méthode à chambre ouverte (Figure 4) et à chambre fermée. Le premier procédé utilise une capsule qui est exposée à l'atmosphère. Il est extrêmement important que les résultats de la PIE ne soient pas assujettis aux changements d'humidité ambiante ou courants d'air [57, 58]. Il est donc souhaitable d'utiliser un système à chambre fermée, réduisant la variabilité de lectures, tout en améliorant la reproductibilité des mesures. Cependant, les instruments en chambre fermée présentent l'inconvénient évident de provoquer une occlusion de la peau. Certains appareils sont équipés d'une chambre de condensation pour lutter contre ce problème d'occlusion à l'intérieur de la sonde de mesure, permettant ainsi un enregistrement pendant plusieurs heures. Etant donné que la sécheresse cutanée s'accompagne par une altération de la barrière cutanée, la PIE peut être considérée comme un témoin indirect de l'état d'hydratation. La peau physiologiquement normale se caractérise par des valeurs de PIE inférieures à $5g/m^2.h$.

I.2.4.2 METHODES ELECTRIQUES DE MESURE DE L'HYDRATATION DU *STRATUM CORNEUM*

L'impédance électrique (EI) est définie comme la résistance électrique de la peau à un courant alternatif. La mesure de l'impédance de la peau a été largement étudiée et représente la technique la plus largement utilisée pour évaluer la teneur en eau du SC [42, 44, 59]. Les propriétés électriques de la peau sont directement liées à la teneur en eau du SC [60]. L'augmentation du niveau d'hydratation épidermique se traduit par des changements de propriétés électriques du SC se manifestant par une diminution de la capacitance, et de l'impédance [61]. La sonde du cornéomètre est faite de deux plaques métalliques placées à proximité, séparées à partir de la surface de la peau par une lame de verre afin d'empêcher la conduction du courant (Figure 5).

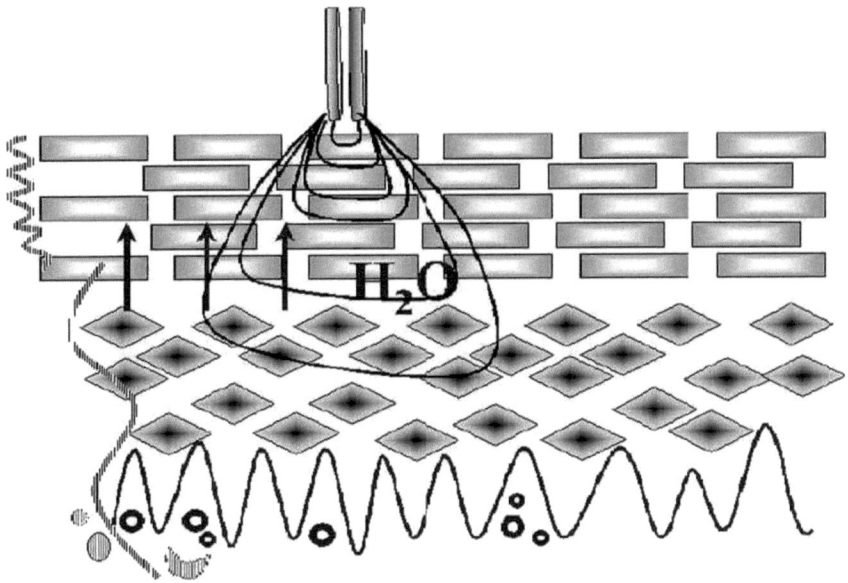

FIGURE 5: DISPOSITIF POUR L'EVALUATION DE L'HYDRATATION UTILISANT LA MESURE D'IMPEDANCE: LA MESURE EST BASEE SUR LA DIFFERENCE DES CONSTANTES DIELECTRIQUES DE L'EAU ET D'AUTRES SUBSTANCES DANS LE CHAMP DE MESURE ELECTRIQUE. LA FAIBLE FREQUENCE DE FONCTIONNEMENT (40 A 75 HZ) DE LA SONDE EST SENSIBLE AUX CHANGEMENTS RELATIFS DE LA CONSTANTE DIELECTRIQUE DU SC [55].

L'inconvénient le plus évident de la mesure d'eau par EI est l'influence de substances autres que l'eau sur l'impédance ou la résistance à l'écoulement [14]. De plus, il a été récemment démontré que la seule corrélation qui existe entre les mesures de EI et les données de la spectroscopie confocale Raman est la teneur en eau dans les couches inférieures du SC [62]. Ces observations semblent être soutenues par les analyses spectrocopiques en proche infrarouge. Celles-ci indiquent que des mesures d'impédance et de capacitance sont relativement insensibles aux petites variations dans la peau sèche [63] et donc peu sensibles à de faibles taux d'hydratation comme ceux des parties superficielles du SC (~15%).

On observe donc un intérêt grandissant à l'utilisation des techniques spectroscopiques telles que la spectroscopie Raman ou infrarouge dans la mesure de l'hydratation, plutôt que des mesures électriques

I.2.4.3   SPECTROSCOPIE RAMAN CONFOCALE ET HYDRATATION CUTANEE

Les techniques « standards » actuellement utilisées pour déterminer la teneur en eau et la sévérité de l'état xérotique ne fournissent aucune information quantitative en ce qui concerne la répartition effective de l'eau dans le SC, son épaisseur ou la composition moléculaire de la peau. Cependant il est indispensable que les techniques choisies pour l'analyse moléculaire de la peau soient non invasives, conduisant à un essor considérable de l'utilisation de celles impliquant les spectroscopies vibrationnelles [64]. Ces méthodes permettent à la fois des analyses *ex vivo* et *in vivo*. En effet, les techniques spectrales donnent autant d'informations sur l'organisation moléculaire du SC en termes d'eau, de lipides ou NMF *in vivo* et *ex vivo*. Elles permettent une localisation spatiale à l'échelle   micrométrique,   la   caractérisation   moléculaire,   organisationnelle   et fonctionnelle des entités tissulaires étudiées.

La spectroscopie Raman peut être utilisée pour suivre les profils de libération de substances dans la peau et permet de déterminer l'épaisseur du SC *in vivo*. La micro-spectroscopie confocale Raman a permis de mesurer avec précision les profils de concentration de l'eau et des NMF en profondeur de la peau humaine [5]. Par la suite, plusieurs travaux ont décrit la mesure des gradients de l'eau et d'autres substances dans le SC [1, 65, 66]. En même temps, il a été démontré que la teneur en eau des couches épidermiques basales reste relativement constante, alors qu'une importante discontinuité existe entre l'état d'hydratation des couches épidermiques supérieures [1]. Plus récemment, en plus de confirmer ces observations, Egawa et al. [4] ont également montré que les altérations dans les gradients de concentration de l'eau dépendent de l'âge, du site anatomique et de la saison.

Chrit et al. [48] a décrit les spectrocopies Raman et infrarouge comme outils de dépistage capable d'évaluer les capacités d'hydratation de quelques entités chimiques appliquées par voie topique. De plus, ces techniques spectroscopiques ont été utilisées pour étudier les transitions de phase et l'organisation de chaînes lipidiques *ex vivo* [67-69]. Bouwstra et al. [70] a montré l'importance du contrôle de l'organisation supramoléculaire des lipides à l'intérieur du SC par infrarouge afin de détecter l'action d'émollients lipophiles capables d'interagir avec ces lipides [71]. Les spectroscopies Raman et infrarouge permettent non seulement l'identification des échantillons analysés au niveau moléculaire, mais également des informations détaillées portant sur la l'organisation supramoléculaire de la peau peuvent être obtenus sur base des positions, des intensités relatives et des formes des bandes du spectre Raman [67, 72, 73] caractéristiques d'une structure (structure secondaire des protéines, organisation supramoléculaire des lipides ...) et d'un environnement moléculaire (interactions intra- et / ou inter moléculaires) bien déterminé. Gniadecka et al. [74] ont utilisé la spectroscopie Raman pour mieux comprendre les altérations liées à l'âge dans la structure des protéines et de l'eau. Leur étude a révélé un haut degré de repliement des protéines et une augmentation du contenu de l'eau tétraèdre (eau non-liée) dans la peau photo-vieillie. Toutefois, cette technique reste difficile à mettre en œuvre en raison d'un signal faible et des phénomènes de fluorescence qui parasitent le signal Raman.

Ces travaux donnent une illustration de la puissance des spectroscopies vibrationnelles, Raman et infrarouge pour mieux comprendre le mode d'action des différents agents hydratants présents dans le SC à l'échelle moléculaire. Cependant, jusqu'à ce jour, aucun travail n'a été réalisé sur un SC isolé dans la perspective d'établir une relation claire entre le taux d'hydratation et les changements dans les descripteurs de conformation et d'organisation de lipides ou de la structure secondaire des protéines dans le SC.

I.2.4.4   MESURE DES PROPRIETES MECANIQUES DU *STRATUM CORNEUM*

L'hydratation influence considérablement les propriétés mécaniques du SC à la fois *in vivo* et *in vitro* notamment, l'allongement à la rupture ou le module de Young qui peuvent être modifiés jusqu'à dix fois selon les conditions expérimentales. *In vivo*, le SC et les autres couches de l'épiderme sont hydratés par le derme sous-jacent. Par conséquent, la mesure *ex vivo* est très attrayante à travers la possibilité d'isoler et de contrôler précisément la couche de la peau à étudier. Ainsi, l'effet des conditions environnementales sur les propriétés de traction uni-axiales du SC a pu être analysé *ex vivo* par plusieurs auteurs depuis les années 1960 [75, 76]. Etant donné que le SC est extrêmement sensible à son environnement, un contrôle strict de l'environnement des mesures est fondamental. Les propriétés mécaniques telles que la limite d'élasticité, allongement à la rupture et le module de Young ont été mesurées. La limite d'élasticité correspond à la contrainte à partir de laquelle l'échantillon cesse de se déformer d'une manière réversible (élastique) c'est à dire qu'une fois relâché, il ne retrouve plus ses dimensions initiales. L'allongement à la rupture est le pourcentage d'élongation avant cassure d'un échantillon soumis à une traction. Le module de Young se défini comme la constante qui relie la contrainte et la déformation pour des déplacements inférieurs à la limite d'élasticité. Ces quelques paramètres constituent les principales caractéristiques des matériaux relativement faciles à mesurer. En plus de la traction, différents auteurs ont effectués des essais de nano-indentation sur SC isolé [77] et sur le-stripping [78] et ont obtenue des résultats semblables mais avec différents ordres de grandeur.

Dans les années 1970, différents chercheurs comme Wildnauer et al. [79], Agache et al. [80] ont étudié l'effet de l'HR sur la force de traction et la courbe d'allongement du SC humain isolé. Ils ont constaté que pour le SC totalement hydraté à température ambiante, trois différentes régions peuvent être distinguées (Figure 6). La première région (phase I) est la région Hooke définissant une extension réversible de 0 jusqu'à la limite d'élasticité, généralement atteinte pour des déformations jusqu'à 10%. La deuxième région (phase II) correspond à des déformations plastiques allant de 20% à 125%. Pour un échantillon tiré jusque dans la phase plastique, la courbe de décharge a la même pente que la courbe de traction et atteste une déformation résiduelle. Dans la troisième région (phase III), une phase de durcissement (également irréversible) du matériau apparaît. Il est notamment associé à des modifications à l'échelle moléculaire et intervient jusqu'à environ 190 % de déformation, correspondant à la rupture effective de l'échantillon. La même expérience a été réalisée pour des éprouvettes de SC à 32% HR et 76% HR. A basse HR, la phase plastique n'a pas été observée et la fracture s'est produite à 22% d'allongement. A 76% d'HR, la fracture s'est produite à 103%, correspondant à environ la moitié de l'allongement d'un SC totalement hydraté, et le point de plasticité était à environ 30% d'allongement. Les auteurs ont conclu que la phase plastique n'a lieu qu'à HR> 70%. Cette valeur a également été constatée par Papir et al. [81]. Enfin, La phase d'endommagement peut être observée uniquement à près de 100% HR. Cette conclusion suggère que c'est seulement dans des conditions de forte hydratation que le SC affiche un comportement hyperélastique caractéristique des matériaux cahoutchouteux [79]. Ce comportement non linéaire a été également observées par les études *in vivo* qui ont suivies [82, 83].

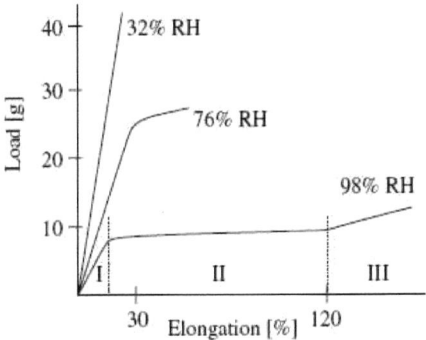

FIGURE 6: COURBES TYPIQUES DE FORCES D'ALLONGEMENT POUR DIFFERENTES NIVEAU D'HUMIDITE RELATIVE MONTRANT LES DIFFERENTES PHASES. OBTENU A PARTIR DE WILDNAUER ET AL.[79].

D'autres études se sont intéressées à l'effet de l'HR sur le module d'élasticité, la pente initiale de la zone I de la courbe contrainte-déformation [81, 84, 85]. Les expériences de traction ont été effectuées pour des HR allant de 26 à 100% à 25 °C dans un environnement de température et d'humidité contrôlées. Autour de 30% d'HR les modules de Young (E) variaient entre 2000MPa et 8900MPa, tandis que pour 100 % d'HR cette valeur varie entre 5MPa et 12MPa selon l'étude [81, 84, 85].

Quelques années plus tard, Nicolopoulos et al. [86] ont constaté une valeur plus élevée pour le module de Young alors rapporté par Wildnauer et Papir. La raideur des échantillons à 85% d'HR a été calculée à partir de la partie terminale de la courbe de déformation et équivalait à 0,54 GPa. Leurs échantillons étaient plus courts et plus larges que ceux utilisés par d'autres auteurs. Ils ont observé que le pré-conditionnement réduit l'énergie de rupture de plus de deux tiers à une humidité élevée.

Outre l'effet de l'HR Van Duzee et al. [87] ont étudié l'influence de la température sur les propriétés mécaniques du SC. Les résultats ont montré que pour chaque HR (30% - 95%), le module de Young mesuré est presque invariant entre 2 ° C à 45 ° C de température). Ils ont conclu que le SC peut être considéré comme un polymère en fonction de ses propriétés mécaniques, car le module dépend de la teneur en eau du SC, pouvant être modifié par liaison de petites molécules et est considéré indépendant de la température jusqu'à un certain seuil.

Il était donc supposé que la force appliquée dépend principalement des protéines (à la fois la kératine et les autres protéines cornéocytaires) [84]. Dans la région de Hooke les liens intermoléculaires sont tendus élastiquement et brisées dans la région plastique. L'eau est censée agir comme un plastifiant provoquant une transition d'un polymère d'un état vitreux (qui se comporte comme un corps solide élastique) à un état hyperélastique (avec un comportement non linéaire). A basse HR, les grands mouvements de chaînes sont limités et l'extension se fait par étirement des liaisons. Aux HR élevées, les liaisons hydrogènes sont hydratées (et donc affaiblies), mais les liaisons disulfures restent intactes, ce qui entraîne un réseau légèrement réticulé. De telles caractéristiques se retrouvent dans d'autres tissus kératinisés comme le cheveu.

Ces auteurs décrivent la relation entre la capacité de rétention d'eau et les propriétés mécaniques du SC. Ils ont observé que l'extraction des lipides du SC avec les solvants organiques, suivie d'une extraction par l'eau, réduit le pouvoir de rétention d'eau et augmente le module d'élasticité [84, 88]. Le traitement par des solvants lipidiques cause la rupture de la membrane cornéocytaire qui maintient les substances hydrosolubles dans les cellules. L'extraction aqueuse suivante enlève ensuite ces substances hydrosolubles. Les auteurs suggèrent que lors de l'extraction de ces substances hydrosolubles, le réseau de protéines s'effondre conduisant en plus de liaisons protéine-protéine dans la cellule à la formation d'une structure plus compacte avec de pouvoirs amoindris de rétention d'eau, et un module d'élasticité plus élevé. Toutes ces suppositions et déductions doivent être vérifiées par des méthodes appropriées. Nous mettrons ainsi en œuvre les spectroscopies vibrationnelles afin de pouvoir analyser les modifications supramoléculaires liées à ces propriétés mécaniques.

A part les mesures de contraintes de traction uni-axiales, des mesures de traction multi-axiales du SC ont été réalisées. Récemment, l'équipe du Pr. R. Dauskardt [3, 89] s'est inspirée d'une technique de mesure de tension des couches minces constitués de métaux [90-92] et l'a adaptée pour mesurer le changement des propriétés mécaniques du SC isolé en fonction de son niveau d'hydratation.

Le SC hydraté est placé sur une lame de verre de borosilicate et une chambre à conditions environnementales contrôlées est ensuite utilisée afin d'analyser l'influence de l'humidité et de la température sur la contrainte mesurée. Lors de la déshydratation, le SC, solidaire de la lame, se rétracte créant une courbure de cette dernière. L'état de tension au sein du système est donc déterminé par l'évaluation du rayon de courbure du substrat et en tenant compte de ses propriétés mécaniques. Ces auteurs ont également montré que l'analyse a permis la discrimination du pouvoir hydratant de différents émollients et une prédiction de l'apparition de fissures [3, 89, 93, 94].

En combinant toutes ces observations, il est évident que les propriétés mécaniques du SC sont très dépendants des conditions environnementales. Toutefois, les valeurs des tensions à la rupture et les modules de Young mesurés par différents auteurs varient considérablement. Les causes possibles sont le pré-conditionnement des échantillons, les variations des conditions environnementales et les écarts dans la mesure de force.

En effet, différents paramètres comme la vitesse de déformation ou la direction de la traction influencent les mesures mécaniques [95]. Un certain nombre d'auteurs a montré qu'avec l'augmentation de la vitesse de déformation, le module d'élasticité ainsi que la contrainte à la rupture augmentent alors que la déformation à la rupture diminue [96, 97]. Néanmoins, il existe une valeur seuil (variable selon les auteurs) au delà de laquelle la vitesse de déformation influence peu les mesures [98], on parle dans ce cas de déformations « quasi-statiques »..

Les inconvénients majeurs des expériences *in vitro* sont les effets inconnus des méthodes de préparation de l'échantillon sur les propriétés mécaniques. Il est donc évident que la précision des résultats doit être améliorée afin d'être en mesure de prédire le comportement mécanique du SC. Ceci passera par la standardisation des protocoles d'analyse, et sera complété par la compréhension des processus moléculaires liés au comportement mécanique.

Cependant, ces essais réalisés *ex vivo* ne peuvent pas être adaptés à des analyses *in vivo*. Outre les difficultés technologiques, l'incapacité d'exclure l'influence des tissus sous-jacents constitue une question prépondérante. Du point de vue de la déshydratation, le stress induit par la chambre de contrôle de l'environnement serait immédiatement compensé par un apport d'eau dermique et épidermique.

L'analyse *in vivo* des propriétés mécaniques du SC reste donc à réaliser. Différentes méthodes sont proposées pour les mesurer : l'indentation (mesure de la déformation de la peau sous pression) [78], levatometrie (mesure du relâchement de la peau en tirant verticalement) [99], ballistometrie (mesure les propriétés mécaniques par l'impact d'une masse avec peau : l'élasticité, la rigidité, la vitesse de rebond etc...), la torsion et la succion (mesure de l'élévation de la peau causée par une dépression dans une chambre cylindrique) [82, 100]. La distinction entre la contribution du SC et celle des couches sous-jacentes (épiderme, derme) reste néanmoins un problème dans l'utilisation de ces techniques *in vivo* [101].

Pour toutes ces raisons, nous avons choisi de mettre en œuvre des protocoles bien définis des modèles *ex vivo* de stress induit de déshydratation / réhydratation du SC isolé qui permettent la détection des variations biomécaniques et physico-chimiques décrites *in vivo*.

CHAP II. PRINCIPALES TECHNIQUES ANALYTIQUES MISES EN ŒUVRE DANS L'ETUDE TISSULAIRE DU *STRATUM CORNEUM*.

II.1.   LES SPECTROSCOPIES VIBRATIONELLES : PRINCIPE ET INSTRUMENTATION

Les spectroscopies optiques constituent l'ensemble des méthodes qui mesurent les phénomènes d'interaction lumière/ matière et utilisent donc la lumière pour l'étude des processus moléculaires. Les positions relatives des atomes dans une molécule ne sont pas exactement fixées mais fluctuent sans cesse en raison de divers types de vibrations. Les spectroscopies vibrationnelles (diffusion Raman et absorption infrarouge) sont des techniques de spectroscopie optique basées sur des transitions entre des niveaux vibrationnels du même état électronique. Les théories quantiques et mécaniques de ces techniques Raman et infrarouge sont très bien décrites dans la littérature [102-106]. Notons simplement que la diffusion Raman est active lorsqu'il y a un changement de la polarisabilité de la molécule, tandis que l'absorption infrarouge exige une modification du moment dipolaire. Ces techniques permettent de mesurer l'interaction des radiations électromagnétiques incidentes avec les vibrations moléculaires spécifiques de l'échantillon ; telles que les élongations ou les déformations angulaires des liaisons (Figure 7). Les deux techniques sont donc complémentaires et permettent d'obtenir une signature vibrationnelle complète de l'échantillon.

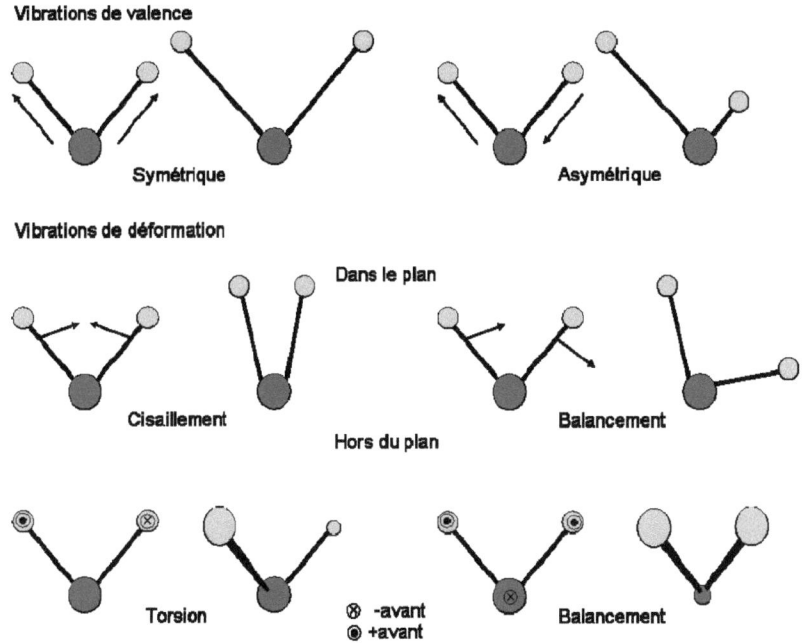

Vibrations de valence

Symétrique — Asymétrique

Vibrations de déformation

Dans le plan

Cisaillement — Balancement

Hors du plan

Torsion — Balancement

⊗ -avant
◉ +avant

FIGURE 7: MODES DE VIBRATION D'UN GROUPEMENT PLAN CH$_2$: LES SCHEMAS DU DESSUS REPRESENTENT LES VIBRATIONS D 'ELONGATION SYMETRIQUE (GAUCHE) ET ASYMETRIQUE (DROITE). LES SCHEMAS DU MILIEU REPRESENTENT LES VIBRATIONS DE DEFORMATION, DANS LE PLAN, DE CISAILLEMENT (GAUCHE) ET DE BALANCEMENT (DROITE). ENFIN, LES SCHEMAS DU BAS REPRESENTENT LES VIBRATIONS DE DEFORMATION, HORS DU PLAN, DE TORSION (GAUCHE) ET DE BALANCEMENT (DROITE).

Il est donc possible de déduire d'un spectre des informations sur la nature et la structure des molécules, sous leur forme libre ou liée, et se présente comme une empreinte digitale des mélanges complexes permettant ainsi leur caractérisation.

II.1.1   NOTIONS DE BASE DE LA SPECTROSCOPIE RAMAN

II.1.1.1   PRINCIPE DE L'EFFET RAMAN

En spectroscopie Raman, l'analyse se fait par excitation de l'échantillon avec une onde électromagnétique de fréquence ν_{exc} [107-112]. De cette interaction onde-matière en résultent deux types de radiations. La première très majoritaire correspond à la diffusion Rayleigh: la radiation incidente est diffusée élastiquement sans changement d'énergie donc de longueur d'onde (environ 1 photon sur 10^4). La deuxième correspond à la diffusion Raman: des photons dans un nombre très limité de cas (environ 1 photon sur 10^8) gagnent (ou cèdent) de l'énergie aux photons incidents produisant ainsi les radiations Stokes avec une fréquence $\nu_{exc} - \nu_x$; relatives à un passage du niveau d'énergie ν_0 à ν_x (ou anti-Stokes avec une fréquence $\nu_{exc} + \nu_x$; relatives à un passage du niveau d'énergie ν_x à ν_0) (Figure 8) ; x représentant les différents niveaux vibrationnels d'un état électronique donné. Il peut se produire une diffusion Raman de résonance quand la diffusion est accompagnée de transitions électroniques moléculaires. Cela demande que les longueurs d'onde de l'excitatrice et de l'absorption électronique de la molécule analysée soient proches.

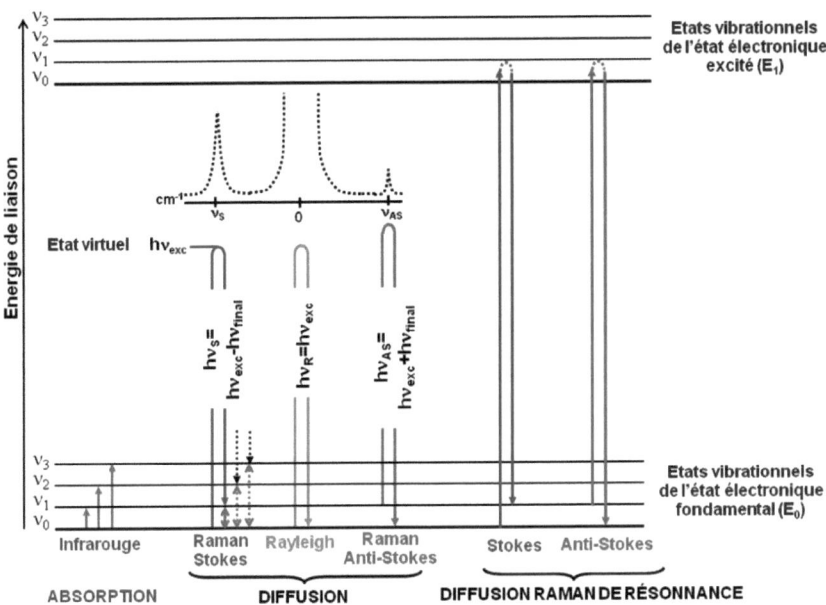

FIGURE 8: DIAGRAMME DE JABLONSKY : EVOLUTION DE L'ENERGIE DE VIBRATION RESULTANT DE L'INTERACTION DE LA LUMIERE AVEC LA MATIERE. ADAPTE DE [113]

Si on suppose le système en équilibre thermique, alors la répartition des populations N_{v1} et N_{v0} des niveaux v_1 et v_0 est définie par la loi de Boltzmann

$$\frac{N_{v1}}{N_{v0}} = \exp(-\frac{\Delta E}{kT})$$

ÉQUATION 2

Où $\Delta E = E_1 - E_0 = hv$

h : constante de Planck ($6{,}62606957 \times 10^{-34}$ m² kg s⁻¹)

v : fréquence (unité : s⁻¹=Hz)

k : constante de Boltzmann ($1{,}3806488 \times 10^{-23}$ m² kg s⁻² K⁻¹)

T : température (unité : K)

On en déduit que les niveaux energétiques inférieurs sont les plus peuplés à température ambiante (300K). L'effet Stokes part de l'état vibrationnel le plus stable du niveau électronique fondamental i.e. état le moins énergétique v_0, donc le plus peuplé tandis que l'effet anti-Stokes part d'un état vibrationnel excité v_1 du niveau électronique fondamental qui dans les conditions normales est peu peuplé. Il s'en suit que l'effet

Stokes génère des raies d'intensité plus élevée que celles de l'effet anti-Stokes. De ce fait, en pratique, c'est la diffusion Raman Stokes qui est mesurée. La diffusion Raman se produit lorsque le champ électrique de la lumière excitatrice induit un changement de polarisabilité de la liaison.

D'une part, le diagramme de Jablonsky illustre les échanges d'énergie entre une onde et la matière alors que d'autre part une approche mécanique classique peut expliquer comment évolue les vibrations et sert à interpréter les différentes fréquences du spectre.

Considérons une onde monochromatique de fréquence v_{exc} rencontrant des molécules de la matière. Le champ électrique de cette onde se met sous la forme [107]:

$$E = E_0 \cos(2\pi v_{exc} t)$$
ÉQUATION 3

Où E_0 est l'amplitude de l'onde. L'interaction entre le champ électrique et le nuage électronique de l'échantillon va créer un moment dipolaire induit μ défini par

$$\mu = \alpha E$$
ÉQUATION 4

Où α est la polarisabilité de l'échantillon. Pour que l'échantillon donne lieu à un effet Raman, la polarisabilité de l'échantillon doit être de la forme

$$\alpha = \alpha_0 + (r - r_{\acute{e}q})\left|\frac{\partial \alpha}{\partial r}\right|$$
ÉQUATION 5

Où α_0 est la polarisabilité de l'échantillon à la distance internucléaire d'équilibre $r_{\acute{e}q}$, et r la distance internucléaire à tout instant.

Aussi $r - r_{\acute{e}q} = r_m \cos(2\pi v t)$ où r_m est la séparation nucléaire maximale par rapport à la position d'équilibre. En substituant les deux dernières relations dans la première et en appliquant la formule de trigonométrie :

$$\cos x \cos y = \frac{1}{2}\left(\cos(x+y) + \cos(x-y)\right)$$
ÉQUATION 6

On obtient :

$$\mu = \alpha_0 E_0 \cos(2\pi v_{exc} t) + \frac{E_0}{2} r_m \left|\frac{\partial \alpha}{\partial r}\right| \cos[2\pi(v_{exc} - v)t] + \frac{E_0}{2} r_m \left|\frac{\partial \alpha}{\partial r}\right| \cos[2\pi(v_{exc} + v)t]$$
ÉQUATION 7

Le premier terme de cette équation représente la raie Rayleigh, alors que les deux derniers représentent respectivement les raies Stokes et Anti-Stokes.

II.1.1.2   SPECTRE RAMAN

La variation d'énergie observée sur le photon nous renseigne alors sur les niveaux énergétiques de vibration des liaisons atomiques concernées. Puisque la molécule comporte plusieures vibrations et l'échantillon plusieurs molécules, ces informations vibrationnelles sont synthétisées par un spectre Raman. Pratiquement, on affiche le décalage observé entre le nombre d'ondes de la bande Raman et celui de la raie excitatrice (Figure 9). Ceci permet d'obtenir toujours les mêmes positions des bandes caractéristiques d'une molécule sur un spectre Raman quelle que soit la longueur d'onde du laser utilisée. Le zéro de l'échelle en nombre d'onde correspond à la position de la raie excitatrice (ν_{exc}). La définition d'un spectre Raman peut donc se formuler comme suit :

Définition : Un spectre Raman représente l'évolution de l'intensité des ondes électromagnétiques diffusées par la matière excité par une source laser en fonction de la différence entre les nombres d'ondes des photons incidents et ceux des photons diffusés. Leurs intensités respectives s'expriment sous la forme suivante [108] :

$$I_{Stokes} = k \, (\nu_{exc} - \nu)^4 \, I_{exc} \qquad\qquad \text{ÉQUATION 8}$$

$$I_{Anti\text{-}stokes} = k \, (\nu_{exc} + \nu)^4 \, I_{exc} \qquad\qquad \text{ÉQUATION 9}$$

Avec $k = \exp(-\dfrac{\Delta E}{kT})$: constante à une température donnée

ν_{exc} : fréquence de la radiation incidente

ν : fréquence de vibration de la liaison moléculaire excitée

I_{exc} : intensité de la radiation incidente

Les spectres Raman s'expriment en fonction du nombre d'onde inversement proportionnel à la longueur d'onde : $\overline{\nu} = \dfrac{1}{\lambda}$

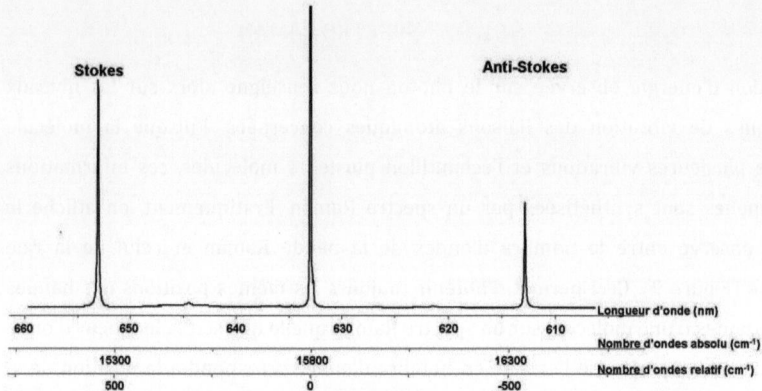

FIGURE 9: SPECTRE RAMAN DU SILICIUM (LONGUEUR D'ONDE D'EXCITATION 632,8 NM) [114].

<u>**Composition du signal Raman enregistré**</u>: Outre le signal Raman proprement dit de l'échantillon étudié, un spectre enregistré contient d'autres signaux. Parmi les signaux parasites qui peuvent accompagner les spectres Raman, on compte :

➢ Un fond de fluorescence ; souvent appelé « fluorescence-like » : il correspond à la fluorescence parasite de l'échantillon.
➢ Le signal Raman du support sur lequel repose l'échantillon.
➢ Des phénomènes instrumentaux comme:
   o le courant noir (signal collecté par le détecteur en absence de toute illumination);
   o spikes (signaux électriques du CCD)

Remarque : les phénomènes qui parasitent un spectre Raman sont éliminés par des prétraitements ou par l'adaptation des conditions d'enregistrement.

Les supports de mesure : Dans la plupart des expérimentations de spectroscopie Raman, l'échantillon à analyser est placé sur un support dont la nature diffère d'une application à l'autre. Pour que le spectre du support n'intefère pas avec le spectre Raman de l'échantillon, ce support sera choisi pour sa faible activité Raman, ou alors, le cas échéant, pour son activité Raman localisée dans des bandes spectrales différentes de celles du spectre d'intérêt. Souvent les acquisitions de spectres sont réalisées sur des supports (CaF_2, BaF_2) ne comportant des signaux qu'à l'extérieur de la gamme spectrale d'intérêt pour les molécules biologiques.

Tous les microspectromètres Raman sont composés principalement (Figure 10) :

- d'une source excitatrice (laser),

- d'un microscope de focalisation et de collecte de la lumière diffusée,

- d'un système de filtrage,

- d'un système de séparation des longueurs d'ondes,

- d'un système de détection.

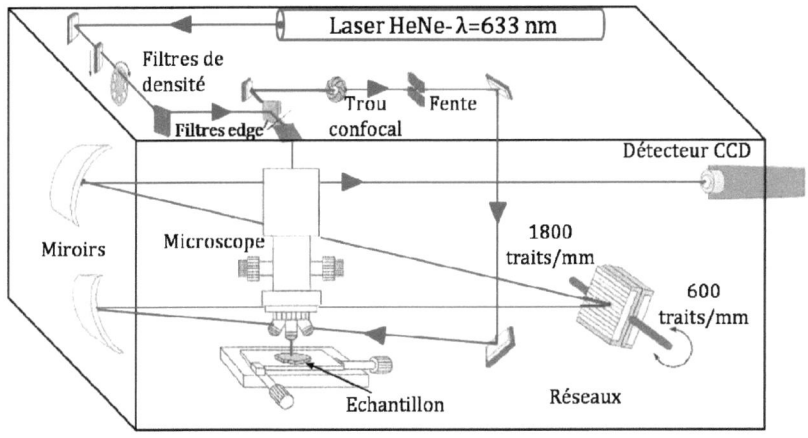

FIGURE 10: SCHEMAS DU MICROSPECTROMETRE RAMAN LABRAM

1. Source

On distingue essentiellement les lasers continus et les lasers pulsés. Le choix de la longueur d'onde d'excitation résulte d'un compromis : plus l'onde est énergétique (longueur d'onde basse ou fréquence élevée), plus la diffusion est intense, mais plus le risque d'induire de la fluorescence parasite est important notamment pour les échantillons biologiques.

Les principaux types de lasers utilisés en spectroscopie Raman sont répertoriés dans le *tableau 1*.

TABLEAU 1: CARACTERISTIQUES DE PRINCIPAUX LASERS UTILISEES EN SPECTROSCOPIE RAMAN

Laser, nm	Type	Gamme de longueur d'onde, nm (stokes, 100-3000 cm^{-1})	Risque de fluorescence	Remarques
Lasers visibles				
488	Ar+	490-572	Très élevé	Laser modulable
514,5	Ar+	517-608	Très élevé	Bonne sensibilité
532	Doublé (Nd :YAG)	535-633	Elevé	Meilleure sensibilité
633	He-Ne	637-781	Assez élevé	Bonne sensibilité
660	Pompé Nd:YLF	664-823	Moyen	Meilleur compromis sensibilité/ fluorescence
Lasers proche infrarouge				
780-785	Diode	791-1027	Moyen	Limité à 3000 cm^{-1} pour détecteur Charged Coupled Device (CCD)
830	Diode	837-980	Très faible	Limité à 2000 Cm^{-1} pour détecteur CCD
1064	Nd :YAG (pulsé)	1075-1563	Nul	Peu sensible, Raman à transformée de Fourier

2. Système de collecte

Il sert à collecter la lumière diffusée pour l'envoyer vers le spectrographe. Différents systèmes de collection sont utilisés, comprenant des microscopes, des fibres optiques. Ils peuvent comporter des chambres d'acquisition et être équipés de platines motorisées. Des sondes confocales très compactes et portables, pouvant être utilisés pour des analyses *in vivo,* sont développées. Nous détaillerons leur structure plus loin dans ce manuscrit.

3. Filtrage

Un filtre passe-bande (interférentiel) est placé après la source afin de filtrer les raies parasites. Un laser de puissance modulable, et un jeu de filtres atténuateurs, permet de réduire l'intensité du rayonnement incident, ce qui évite dans certains cas une dégradation photothermique de l'échantillon. L'utilisation de filtres Notch (élimination d'une raie spécifique) ou des filtres Edge (filtres passe-haut) permet d'exclure, par réflexion, la zone spectrale contenant la raie excitatrice du laser et la diffusion Rayleigh (beaucoup plus intense que la diffusion Raman).

4. Séparation des longueurs d'ondes

La séparation des longueurs d'ondes peut être réalisée à l'aide d'un système dispersif (réseaux à diffraction) ou d'un interféromètre (pour la spectroscopie Raman à transformée de Fourier « Raman-TF », plutôt utilisée pour les excitations dans le proche infrarouge).

5. Détecteurs

Les détecteurs multicanaux de type CCD (*Charged Couple Device*) sont les plus communément utilisés avec les systèmes dispersifs tandis que les détecteurs InGaAs (Indium-Gallium-Arsenide) sont plutôt réservés à la spectroscopie Raman-TF.

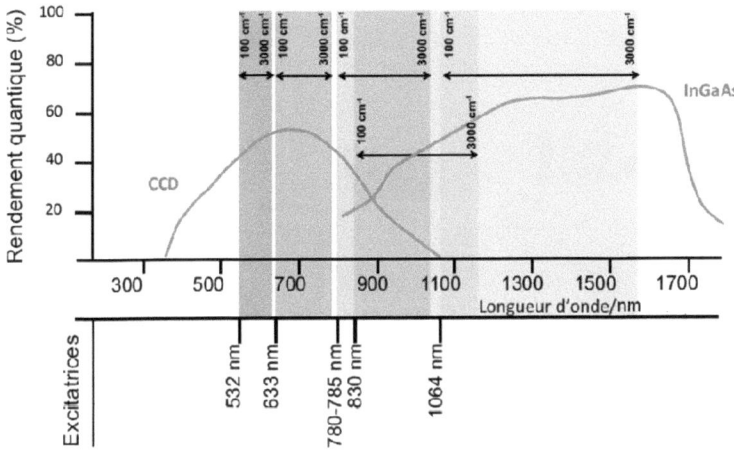

FIGURE 11: RENDEMENT QUANTIQUE DES DETECTEURS (CCD ET INGAAS) EN FONCTION DE LA LONGUEUR D'ONDE. ÉTENDUE EN NM DE LA GAMME SPECTRALE (100-3000 CM⁻¹) EN FONCTION DE L'EXCITATRICE [114].

Chaque détecteur a un rendement quantique spécifique (Figure 11) ; la position et l'étendue de la réponse Raman en longueur d'onde sont fonction du laser (*tableau 1*, colonne 3). La réponse optimale est obtenue dans la zone de recouvrement de sensibilité du détecteur et de la réponse Raman. Par exemple, pour un laser à 633 nm, la gamme spectrale 100- 3000 cm-1 correspond à la plage de longueur d'onde 637-781 nm. Sur cette plage de mesure, le rendement quantique d'un détecteur CCD est optimal (Figure 11).

II.1.1.4   MICRO-SONDE RAMAN

Pour des analyses de la peau *in vivo* par microscopie Raman, il existe des appareils comme celui développé par Horiba Scientific (Fig, 12) et River Diagnostic [50, 115]. Le système d'acquisition des données spectrales de ce dernier est constitué d'un microscope inversé figé sur lequel on pose la partie du corps à analyser. Il est donc évident que cette configuration ne permet pas d'analyser différentes zones du corps humain. Ainsi, un système de spectroscopie Raman *in vivo* incluant une micro-sonde confocale facile à manier, a été développé par Horiba Scientific, afin de pouvoir analyser différents sites anatomiques.

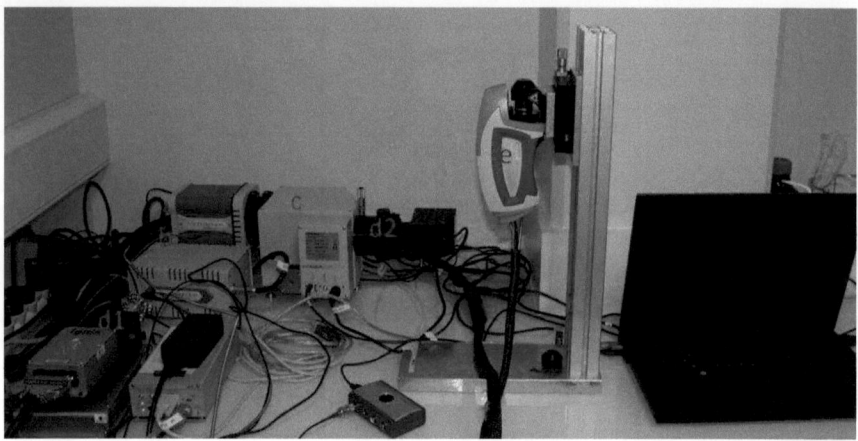

FIGURE 12: SYSTEME MICRO-SONDE RAMAN: (A) SOURCE LASER, (B) CHARGED COUPLED DEVICE (CCD), (C) MICRO HR, (D1 ET D2) SHUTTER LASER ET SHUTTER SPECTROMETRE RESPECTIVEMENT, (F) FIBRE PERMETTANT UNE CONFOCALITE DE 5 µM ET (E) MICRO-SONDE CONFOCALE

Le système comprend :

- Une source laser à 660 nm (Figure 12a). La puissance laser est de 12 mW en sortie d'objectif.

- La micro-sonde confocale (Figure 12e) ; Une fibre optique monomode de 5 µm de diamètre utilisée pour véhiculer la lumière laser, assurant une confocalité de l'ordre de 3 µm (avec un objectif 100X). La sonde comporte également une caméra intégrée permettant de visualiser la surface de l'échantillon.

- Un objectif remplaçable selon les besoins analytiques. Dans nos travaux, nous utilisons un objectif 100X MPlanFLN (NA, WD). Un système piézoélectrique permet un mouvement en Z de l'objectif de ± 100 µm avec des pas d'incrémentation ajustables.

- Un spectromètre Raman, ici de type micro HR (Fig. 12c), équipé de réseaux de différent pouvoir résolutif (400, 600 et 1800 traits/mm) et d'un détecteur CCD (Fig. 12b) de haute sensibilité refroidi à -70°C par effet Peltier.

- Au bout de l'objectif, un embout indépendant de l'objectif et comprenant une ouverture de 1 ou 2 mm, permet d'ajuster la distance entre de la peau et l'objectif de façon à se positionner à la distance de travail optimale. Un système d'autofocus ultrarapide permet une focalisation du laser en surface de la peau, presqu'en temps réel.

Nous avons utilisé ce système pour des analyses directes sur des volontaires permettant ainsi l'étude de la structure et de la santé cutanée dans des conditions réelles.

II.1.1.1   PRETRAITEMENT DES SPECTRES RAMAN

Comme dans toute expérimentation, des paramètres impondérables surviennent lors des enregistrements et polluent les spectres Raman de façon plus ou moins forte. La théorie prédit un ensemble de phénomènes parasites, facilitant ainsi leur élimination. Ces divers phénomènes vont être décrits en même temps que les méthodes pour les éliminer.

Élimination du courant noir :

Le courant noir est le signal mesuré sur les détecteurs CCD dans une obscurité totale, en l'absence de laser excitateur. L'intensité de ce bruit est dépendante de la température du CCD mais également du temps d'exposition des pixels.

Pour s'affranchir de ce courant noir, un spectre est enregistré sans laser ni échantillon, c'est-à-dire qu'aucune lumière n'est dispersée sur le détecteur. Ce spectre du courant noir est donc soustrait des autres spectres enregistrés en chaque point de mesure de l'échantillon à analyser.

Correction de la réponse du détecteur :

L'efficacité du détecteur CCD ainsi que la transmission et/ou la réflexion de la lumière par les éléments optiques du spectromètre dépendent de la longueur d'onde du rayonnement émis. Afin de rendre les mesures indépendantes de l'appareillage, et ainsi uniquement dépendantes de la structure moléculaire de l'échantillon analysé, cette réponse de l'instrumentation doit être corrigée en divisant le signal de l'echantillon par celui de la lumière blanche.

Elimination du bruit :

Le signal Raman se distingue du bruit propre au système de détection utilisé. Pour bien identifier et analyser le signal, il faut filtrer le bruit et éliminer les distorsions éventuelles. Les techniques de lissage souvent utilisées sont : moyenner sur un certain nombre de points (moyenne flottante) et l'algorithme de Savitzky-Golay [116].

Correction de la ligne de base :

Les spectres obtenus lors des acquisitions sont composés essentiellement de deux parties majeures : la luminescence qui se définit comme un phénomène d'absorption et de réémission d'un rayonnement, et le signal Raman qui est la diffusion stokes et anti-stokes du rayonnement incident. L'interférence spectrale avec la luminescence (sous forme de fond continu des spectres) varie d'une mesure à une autre. Raison pour laquelle il est indispensable de l'éliminer avant d'analyser l'effet Raman d'un échantillon. Différentes méthodes ont été jusqu'alors utilisées sans toutefois obtenir une méthode universelle pour toute sorte de données. L'enjeu est de pouvoir mesurer les surfaces et/ou les intensités des pics Raman dépourvues de la luminescence sans déformer les profils des pics ; comme la largeur des bandes. Différents algorithmes permettant d'automatiser de manière reproductible l'élimination de la contribution de la « fluorescence-like » du SC ont été proposés [117].

Normalisation des spectres :

Cette normalisation a pour effet d'accorder à tous les spectres enregistrés la même importance informative, c'est-à-dire s'affranchir de la variabilité de l'intensité absolue lors la comparaison de différents spectres.

II.1.2   NOTIONS DE BASE DE LA SPECTROSCOPIE INFRAROUGE

II.1.2.1   PRINCIPE DE L'ABSORPTION INFRAROUGE

Nous avons vu que pour obtenir un effet Raman, une modification de la polarisabilité (moment dioplaire induit) de la molécule est nécessaire. L'absorption infrarouge est pour sa part reliée principalement aux modifications du moment dipolaire permanent (μ_p) de la molécule, lors de son mouvement vibrationnel ou rotationnel [118-123].

Le moment dipolaire permanent autour de la position d'équilibre $\mu_{p(0)}$ peut s'exprimer comme suit [119] :

$$\mu_p = \mu_{p(0)} + \left|\frac{\partial \mu_p}{\partial r}\right| r_m \cos(2\pi v t) \qquad \text{ÉQUATION 10}$$

Où

r : la distance internucléaire à tout instant.

r_m: la séparation nucléaire maximale par rapport à la position d'équilibre.

v : la fréquence de vibration de la liaison moléculaire

Pour que l'absorption infrarouge aie lieu, il faut que $\left|\frac{\partial \mu_p}{\partial r}\right|$ soit différent de zéro

Dans ces conditions, le rayonnement électromagnétique dont la fréquence correspond à celle de la fréquence naturelle de vibration de la molécule transfère son énergie à celle-ci conduisant ainsi à une augmentation de l'amplitude de vibration de la molécule. On enregistrera une diminution de l'intensité réfléchie ou transmise. La région infrarouge du spectre comprend la région de longueurs d'onde comprises entre environ 0,78 et 1000 µm (ou de nombre d'onde de 12 800 à 10 cm^{-1}). Il est utile, tant sur le plan des applications que sur le plan de l'instrumentation, de diviser cette zone du spectre en trois régions :

L'infrarouge proche : 12 800 à 4000 cm^{-1}

L'infrarouge moyen : 4000 à 400 cm^{-1}

L'infrarouge lointain : 200 à 10 cm^{-1}

Le domaine moyen infrarouge correspond au domaine d'énergie de vibration des molécules (2,5 µm < λ < 25 µm).

Toutes les vibrations ne donnent pas lieu à une absorption, cela va dépendre aussi de la géométrie de la molécule et en particulier de sa symétrie. A titre d'exemple, les molécules homopolaires telles que O_2, N_2, Cl_2 n'absorbent pas en infrarouge parce que elles ne subissent pas de variations de moments dipolaires lors de leur vibrations.

La position des bandes d'absorption va dépendre en particulier de la différence d'électronégativité des atomes et de leur masse. Par conséquent à une substance de composition chimique et de structure donnée va correspondre un ensemble de bandes d'absorption caractéristiques permettant son identification. Il suffit par la suite d'appliquer la loi de Beer-Lambert à une bande spectrale infrarouge caractéristique d'un groupement fonctionnel de la molécule pour la quantifier.

$$I = I_0 e^{(-\varepsilon l c)}$$

ÉQUATION 11

Où

I_0 : Intensité incidente,

I : Intensité transmise,

ε : Coefficient d'extinction moléculaire,

c : Concentration de l'échantillon,

l   : Longueur traversée par le faisceau.

L'analyse s'effectue à l'aide d'un spectromètre à Transformée de Fourier qui envoie sur l'échantillon un rayonnement infrarouge et mesure les intensités d'absorption des longueurs d'onde auxquelles la substance absorbe.

II.1.2.2   APPAREILLAGE INFRAROUGE

L'appareillage décrit est de type spectromètre infrarouge à transformée de Fourier. Il est constitué de quatre éléments fondamentaux : la source infrarouge, un interféromètre, un détecteur et un système électronique de pilotage et d'affichage des données spectrales récoltées [124-126].

La figure 13 représente le schéma de principe du fonctionnement de l'appareil. Le rayonnement émanant de la source moyen infrarouge est envoyée vers un interféromètre de Michelson, qui représente la partie principale de l'appareil. L'intensité du rayonnement de la source est divisée en deux : 50% retournent à la source et 50% (deux fois 25%) sont recombinés de manière cohérente et envoyés vers l'échantillon. L'intensité des interférences créées par le séparateur de rayons est enregistrée par le détecteur en fonction de la différence de marche induite par le déplacement du miroir mobile de l'interféromètre. La transformée de Fourier de cet interférogramme formé permet d'obtenir un spectre infrarouge.

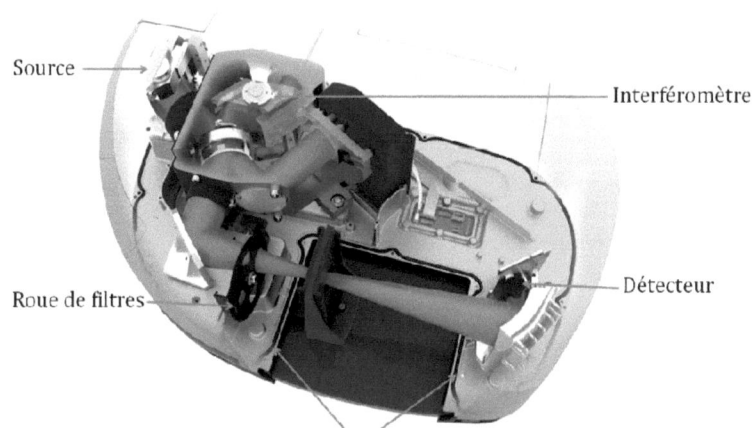

FIGURE 13: SCHEMA DE PRINCIPE DE L'ANALYSE PAR SPECTROSCOPIE D'ABSORPTION INFRAROUGE ; APPAREIL DE TYPE SPECTRUM TWO (PERKIN ELMER)

En fonction des informations recherchées et de la nature de l'échantillon à analyser, il existe plusieurs configurations différentes pour placer l'échantillon entre le faisceau incident et le détecteur. Dans ce travail, nous nous sommes essentiellement servi d'un accessoire appelé "Attenuated Total Reflection" (ATR) permettant des analyses de surface.

II.1.2.3   LA REFLEXION TOTALE ATTENUEE (ATR) :

La technique ATR va concerner la propagation du rayonnement dans un cristal d'indice élevé (n_2) et son absorption-réflexion à l'interface cristal-échantillon d'indice (n_1). La surface optique d'un cristal ATR est le plus souvent fait de séléniure de zinc (ZnSe), de diamant, de silicium (Si) ou de germanium (Ge) avec des indices de réfraction d'environ 2,6 ; 2,4 ; 3,4 ; et 4 et l'indice de réfraction de la matière organique est d'environ 1,5. Pour un angle d'incidence θ supérieur à un angle critique θ_c (sin $\theta_c = n_2/n_1$), la réflexion est totale à l'intérieur du milieu le plus dense qui génère une onde évanescente. Cette onde pénètre de quelques micromètres dans le milieu le moins dense (échantillon) qui se trouve en contact avec le matériau dans lequel se produit la réflexion interne (cristal) [127-129].

La profondeur de pénétration (d_p) est définie comme étant la profondeur à laquelle E_0 l' l'amplitude du champ électrique tombe à une valeur $E = E_0 \exp(-1)$ et est donné par l'équation suivante [130]:

$$d_p = \frac{\lambda}{2\pi n_2 \sqrt{\sin^2 \theta - (n_1 - n_2)^2}} \qquad \text{ÉQUATION 12}$$

La profondeur de pénétration dépend de la longueur d'onde λ.

En pratique, les épaisseurs analysées seront d'autant plus faibles que l'indice de réfraction du cristal, l'angle d'incidence et le nombre d'onde du rayonnement seront élevés. Quoiqu'il en soit, elles ne seront en général pas supérieures à quelques µm. Vu ces faibles profondeurs de pénétration, Il est évident que cette technique est particulièrement adaptée à l'étude du phénomène de surface engendré par la diffusion à travers une membrane.

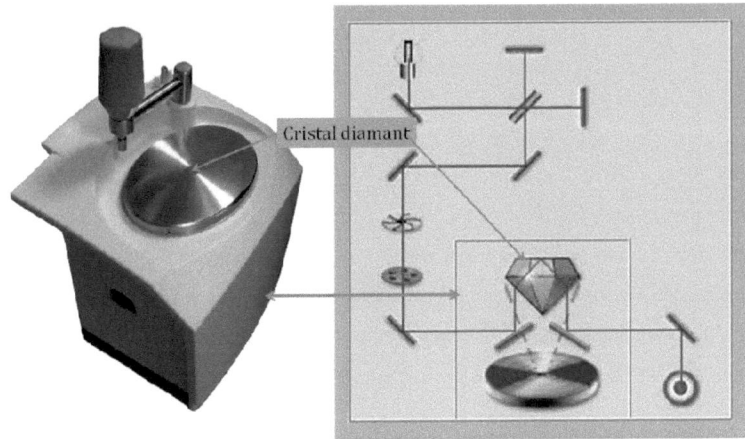

Cristal diamant

FIGURE 14: DISPOSITIF ATR EQUIPE D'UN CRISTAL DIAMANT

La figure 14 représente un dispositif ATR-Diamant équipé d'un bras de pression qui permet d'appliquer un contact contrôlé entre l'échantillon solide et le cristal diamant (dans le cas de cet appareil) et ainsi assurer la reproductibilité des mesures.

II.2. INSTRUMENTS DE MESURES MECANIQUES DU *STRATUM CORNEUM*

II.2.1 MESURES DE COURBURE DU SUBSTRAT : STRESS BI-AXIAL DU *STRATUM CORNEUM*

Comme nous l'avons précédemment décrit, il existe différentes techniques de mesures des propriétés mécaniques. Parmi les techniques utilisées sur le SC ex-vivo, on compte la mesure de la tension du SC par la courbure du substrat, développée par l'équipe du Pr. R. Dauskardt [3, 89] qui se distingue par sa reproductibilité et sa simplicité de mesure.

Cette technique consiste à mesurer la courbure du substrat sur lequel est déposée une couche fine du matériau à étudier. Elle a été largement utilisée pour étudier les contraintes résiduelles dans des dépôts fins de métaux comme l'Aluminium, le Titane etc ... [90-92]. La contrainte des films minces peut être mesurée par la courbure du substrat sur lequel le film collé. Les ordres de grandeurs sont relativement petits, et comme on reste généralement dans la zone de déformation élastique du matériau étudié, la mesure est suffisamment précise pour observer son comportement mécanique.

Rapporté à l'étude du SC, on mesure la courbure du substrat en verre sur lequel l'échantillon à analyser a été préalablement collé. Des lamelles en verre borosilicate, avec des propriétés adaptées à l'analyse, sont choisies comme substrat. Ce substrat est préféré à d'autres supports élastiques en raison de la bonne adhérence du SC. Pour améliorer la réflectivité, des films métalliques (or ou aluminium) sont déposés par évaporation sur un côté du substrat de verre opposé à celui sur lequel le SC est collé.

Un instrument de mesure de courbure du substrat à balayage laser de type Flexus 2020 (Tencor Instruments) est utilisé pour mesurer l'angle de déflexion du substrat, α, en se basant sur les dimensions L_b et L_s en fonction de la position, y, tel que défini dans la figure 15. La valeur de la courbure moyenne est déduite à partir d'une analyse de régression linéaire de α par rapport à la position des mesures [90]. On évalue tout d'abord la courbure initiale avant que le tissu soit déposé sur le substrat pour détecter une courbure résiduelle du substrat qui n'est pas associé au SC. Dans ce qui suit, la courbure représente une variation de courbure par comparaison à la courbure initiale du substrat. La relation entre la contrainte du SC, σ_{sc} et la courbure élastique, K est exprimée avec l'équation de Stoney [131]:

Stress du SC : $\Delta\sigma_{SC} = \left(\dfrac{E_{sub}}{1-v_{sub}}\right)\dfrac{h_{sub}^2}{6\,h_{SC}}\Delta\kappa$ ÉQUATION 13

Courbure du substrat : $\dfrac{1}{R} = \kappa = -\dfrac{1}{2}\dfrac{d\alpha}{dy} = -\dfrac{1}{2}\dfrac{d}{dy}\left(\dfrac{L_s}{L_b}\right)$ ÉQUATION 14

Où E_{sub}, v_{sub}, h_{sub} et h_{sc} sont le module de Young, coefficient de Poisson, épaisseur du substrat et de l'épaisseur de l'échantillon SC, respectivement.

La température et l'HR de l'air dans l'instrument doivent être contrôlées.

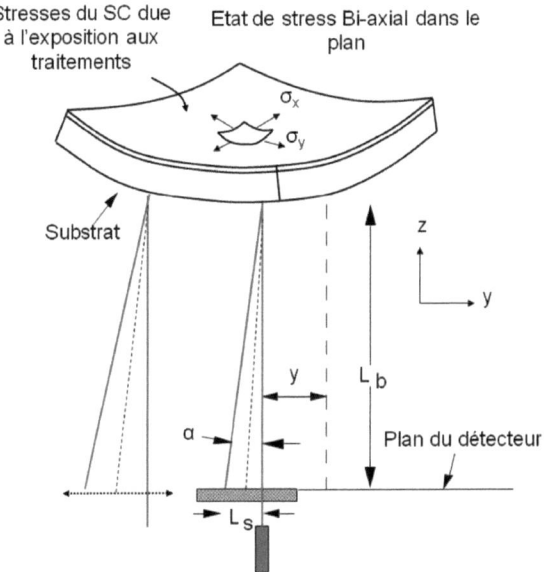

FIGURE 15: DISPOSITIF D'ESSAI POUR LA TECHNIQUE DE LA COURBURE DU SUBSTRAT ILLUSTRANT LE *STRATUM CORNEUM* COLLE SUR UN SUBSTRAT DE VERRE : UN LASER A BALAYAGE EQUIPE D'UN DETECTEUR MESURE L'ANGLE DE DEFLEXION (A) PAR RAPPORT A LA POSITION (Y) SUR LE SUBSTRAT ; LA COURBURE MOYENNE EST CALCULEE A PARTIR D'UNE REGRESSION LINEAIRE DE L'ANGLE DE DEVIATION [3]

II.3.   ANALYSES STATISTIQUES MULTIVARIEES

Les analyses par spectroscopies vibrationnelles combinées aux traitements statistiques multivariés permettent de classer les populations selon leurs différences et d'identifier les critères qui les différencient.

L'analyse de spectres Raman repose sur l'analyse différentielle d'empreintes aboutissant à une expression semi-quantitative de résultats. Souvent, il suffit d'un petit changement dans la structure d'un échantillon pour que son spectre change. Toutefois, ces modifications dans le spectre sont très petites. La difficulté est donc de gérer et de comparer ces empreintes qui contiennent de nombreux signaux correspondant à plusieurs centaines, voir plusieurs milliers de vibrations.

En pratique, il est difficile de comparer visuellement l'ensemble de ces données, d'autant plus, quand le nombre d'échantillons devient important. Ces dernières années, les analyses statistiques **multivariées** ont été mises en œuvre afin d'étudier ou de décrire un ensemble de données spectrales [132]. Elles permettent en effet de synthétiser et de visualiser rapidement une grande quantité d'informations, ceci en projetant les données initiales dans un espace de dimensions réduites qui permet une visualisation aisée de l'ensemble des données [133-136].

Il existe deux types d'analyses statistiques multivariées utilisées dans la littérature dans le cadre de ce genre d'étude [137]:

- Les analyses statistiques **descriptives**, qui ne nécessitent pas d'information « *a priori* » sur la nature des échantillons. Elles permettent de décrire des données et de visualiser la répartition des échantillons. L'analyse en composantes principales (ACP) en est une.
- Les analyses statistiques **explicatives** qui visent à expliquer une ou plusieurs réponses (Y). Dans ce cas, des informations sont données sur la nature des échantillons. Parmi les méthodes explicatives, on citera la régression par moindres carrés partiels (PLS), ainsi que l'analyse discriminante par moindres carrés partiels (PLS-DA).

Ces mêmes méthodes peuvent également être utilisées dans un but prédictif. Un jeu de données initiales est alors utilisé pour construire un modèle explicatif, ce modèle est ensuite appliqué à un nouveau jeu de données qui pourront alors être classées dans les différents groupes.

Les données sont organisées sous forme de matrices. Chaque individu est décrit par un certain nombre de variables X, et parfois par des informations Y. Ainsi, les colonnes des matrices sont constituées des variables X et parfois Y dans le cas d'analyses explicatives. Les lignes qui constituent les matrices correspondent aux individus (échantillons). Les données X, appelées variables correspondent par exemple aux nombres d'ondes en spectroscopie Raman. Dans le cas d'analyses explicatives, les variables Y, appelées observations peuvent être la classe le l'échantillon.

II.3.1   ANALYSES STATISTIQUES DESCRIPTIVES

II.3.1.1   ANALYSE EN COMPOSANTES PRINCIPALES (ACP)

L'ACP permet de représenter les données originelles (individus et variables) dans un espace de dimensions inférieures à l'espace originel, tout en limitant au maximum la perte d'information [138-140]. La représentation des données dans des espaces de faible dimension en facilite, en effet, considérablement l'analyse. Pour cela, un changement de coordonnées est effectué afin de définir de nouveaux axes, appelés composantes principales qui concentreront la plus grande partie de la variabilité des échantillons. Les composantes principales extraites pour décrire la collection de données (variables expérimentales) sont classées en fonction de quantité de variation qu'elles contiennent. La première composante extrait la plus grande source de variance, la seconde composante, orthogonale à la première extrait la plus grand source de variance de ce qui reste à expliquer, et ainsi de suite, si bien que les dernières composantes contiennent surtout le bruit aléatoire. En pratique, les deux ou trois premières composantes permettent souvent d'expliquer plus de 90 % de la variance totale. Les données sont présentées sous forme de « carte factorielle des individus » ou « *score plots*», c'est-à-dire les coordonnées des échantillons dans le nouveau repère généré par l'ACP [141-143].

L'ACP est souvent la première étape dans l'analyse des données afin d'étudier leur répartition. Elle est suivie d'une analyse dite explicative permettant l'interprétation des données afin de déterminer quelles sont les composantes du jeu de données qui contribuent à l'obtention de la répartition observée.

II.3.2   ANALYSES EXPLICATIVES ET DESCRIPTIVES

II.3.2.1   REGRESSION: MOINDRES CARRES PARTIELS (PLS)

La PLS (Partial Least Square ou Projection to Latent Structure) a été développée vers la fin des années 60 par Herman Wold et a évolué jusqu'à sa forme actuelle en 1982 [144-147]. La PLS fournit un modèle qui permet d'expliquer un ou plusieurs observation(s) (Y) en fonction d'un jeu de variables (X) [148-151]. Le fonctionnement des techniques PLS ne sera pas décrit ici en détails. Ce type de modélisation n'utilise pas directement les variables explicatives, mais calcule d'abord les axes PLS sur lesquels sera ensuite réalisée la régression. Ces axes sont des combinaisons linéaires des variables initiales calculées de manière à décrire les plus grandes variations présentes dans le jeu de données et à maximiser la covariance entre les entrées (X) et la sortie (le ou les Y) du modèle. Ces axes sont tous orthogonaux entre eux. En régression PLS, toutes les variables importantes sont conservées et les variables sans importance sont soit exclues, soit participent au modèle, mais avec un faible poids. Les observations Y sont ainsi décrites en fonction des variables X. Pour ce qui est de la visualisation des données, les données sont présentées, tout comme pour l'ACP, sous forme de «carte factorielle des individus», c'est-à-dire les coordonnées des échantillons dans le nouveau repère généré par la PLS. Les données peuvent également être étudiées sous forme de carte des poids factoriels ou « *loading plot* » qui décrivent les relations entre les variables dans le système des composantes.

II.3.2.2   ANALYSE DISCRIMINANTE PAR MOINDRES CARRES PARTIELS (PLS-DA)

L'objectif de la PLS-DA ou analyse discriminante PLS est de construire un modèle qui permet de maximiser la séparation entre les classes auxquelles appartiennent les échantillons [152]. C'est une méthode de régression.

La PLS, aussi bien que la PLS-DA permettent la construction d'un modèle explicatif. Ce modèle permet ainsi d'isoler les vibrations dont l'intensité est caractéristique d'un état organisationnel donné et qui contribuent à la formation des différents groupes que forment les échantillons [153, 154]. Ce modèle construit à partir d'un jeu de données initiales peut également être utilisé pour expliquer un nouveau jeu de données. On cherche ainsi à prédire la classe d'appartenance des individus du nouveau jeu de données.

Cependant les variables d'intérêt mises en évidence par les analyses statistiques multivariées ne sont identifiées que par leur nombre d'ondes dans le cas de la spectroscopie Raman [155, 156]. Il faut donc relier ces vibrations à des conformations moléculaires correspondantes. L'interprétation structurale est ainsi une autre étape de cette analyse

II.4. HUMIDITE RELATIVE

L'humidité relative (HR) est le rapport entre la pression partielle de vapeur d'eau (P_w) contenue dans l'air sur la pression de vapeur saturante (P_{sat}) à une température donnée. Cette pression partielle de vapeur d'eau correspond à la quantité d'eau que contient l'air (humidité absolue) et la pression de vapeur saturante équivaut à la quantité maximale qu'il est capable de contenir (valeur de saturation au-delà de laquelle se produit la condensation).

Elle est exprimée comme suit:

$$HR = \frac{P_w}{P_{sat}} \bullet 100\%$$

ÉQUATION 15

La pression de vapeur d'un liquide pur est la pression exercée par les molécules au dessus du liquide. Pour qu'un liquide se vaporise, l'énergie cinétique des molécules doit être supérieure aux forces intermoléculaires. Un liquide est d'autant plus volatil que sa pression de vapeur est grande.

Sous l'hypothèse que la vapeur se comporte comme un gaz parfait, la pression de vapeur saturante est décrite par l'équation de Clapeyron :

$$\ln\frac{P_{sat}}{P_o} = \frac{M \bullet L_v}{R}\left(\frac{1}{T_0} - \frac{1}{T}\right)$$

ÉQUATION 16

Où :

T_0 : Température d'ébullition de la substance à une pression P_0 donnée, en K

P_0 : Pression atmosphérique, en kPa

P_{sat} : Pression de vapeur saturante, en kPa

M   : Masse molaire de la substance, en kg/mol

Lv   : Chaleur latente de vaporisation de la substance, en J/kg

R   : Constante des gaz parfaits, égale à 8,31447 J/K/mol

T   : température de la vapeur, en K

Comme illustré dans l'équation 16, la pression de vapeur est influencée par la température (Tableau 2).

Tableau 2 : Pression de la vapeur d'eau en fonction de la température

Température (°C)	Pression de vapeur d'eau (kPa)
0	0.61
10	1.23
20	2.34
30	4.24
40	7.37
50	12.33
60	19.92
70	31.18
80	47.34
90	70.11
100	101.33 = 1 atm

La dissolution d'un sel dans l'eau conduit à une diminution de sa pression de vapeur. Pour l'eau pure, les liaisons hydrogènes retiennent les molécules d'eau à l'état liquide. Toutefois, certaines molécules d'eau possèdent suffisamment d'énergie pour passer à l'état gazeux et créer ainsi une pression de vapeur au-dessus du liquide comme décrit précédemment. Dans le cas de solutions saturées de sels, il existe des forces d'attraction supplémentaires aux liaisons déjà présentes pour retenir encore plus fortement le l'eau en phase liquide et diminuer sa pression de vapeur saturante. Cette diminution dépend de la nature du soluté et peut être évaluée par la loi de Raoult :

$$P_{solution} = P_{solvant} \bullet X_{solvant}$$

Où

$P_{solution}$: Pression de la vapeur de la solution

$P_{solvant}$: Pression de la vapeur du solvant pur

$X_{solvant}$: Fraction molaire du solvant

Tableau 3: Influence des sels sur les pressions de vapeur et de HR

Exemple de solutions saturées de sels à 25°C de température	Pression de vapeur d'eau (kPa)	HR
LiCL	0.358	11.3 %
MgCl	1.039	32.8 %
NaCl	2.387	75.3 %
KCl	2.673	84.3 %

Cette propriété permet la préparation de différentes solutions saturées pour générer des taux d'HR bien définis (Tableau 3).

Travaux expérimentaux

CHAP III. OPTIMISATION DES CONDITIONS D'ANALYSE ET DES PARAMETRES D'ACQUISITIONS

III.1. OPTIMISATION DE L'ENVIRONNEMENT D'ANALYSE

L'objectif de cette partie était de mettre en place un système à humidité contrôlée construite dans l'environnement d'analyse du micro-spectromètre Raman pour l'étude des phénomènes d'absorption/désorption de l'eau par le SC.

III.1.1 CONTROLE DE L'HUMIDITE AUTOUR DU MICROSCOPE DU RAMAN

Contexte : Traditionnellement les études de l'hydratation du SC consistait à incuber les échantillons dans les enceintes à humidités bien définies (solutions saturées de sels), et par la suite, ils étaient retirés pour être analysés. En utilisant ce procédé, il a été montré qu'il peut y avoir, en quelques minutes seulement, des pertes (ou gains selon le cas) en eau du SC par échanges avec le milieu. Ceci conduirait donc à des variations de la teneur en eau du SC pendant et entre les acquisitions des mesures.

Cette remarque souligne la nécessité de construire une enceinte à humidité contrôlée autour du microscope afin de réaliser des mesures dans des conditions stables et bien définies.

Différents systèmes ont été envisagés. Parmi eux, une méthode consistant à construire une enceinte hermétique englobant la cloche de sel et la chambre de mesure autour du microscope, tout en accélérant l'équilibre avec un ventilateur. La deuxième méthode consiste à générer une HR contrôlée par mélange d'air sec et d'air humide et l'acheminer dans la chambre des mesures. Les avantages et les limites de ces méthodes seront discutés.

III.1.1.1 CONTROLEUR D'HUMIDITE RELATIVE 01 (RHC-01)

Nous avons construit une chambre à HR contrôlée se composant des éléments suivants :

- Une cloche contenant une solution saturée de sel
- Des sacs plastiques semi-transparents en polyéthylène pour isoler l'ensemble autour du microscope
- Des élastiques pour fixer le sac plastique
- Un hygromètre pour mesurer la température et l'HR à l'intérieur de la chambre

- Un mini-ventilateur pour brasser l'air humide afin d'homogénéiser plus rapidement l'air dans la chambre.

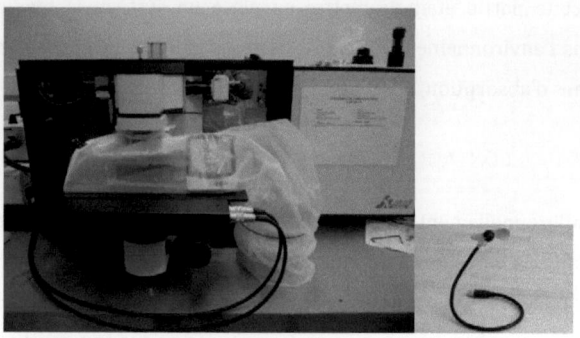

FIGURE 16: RELATIVE HUMIDITY CONTROLER 01 (RHC_01)

5 sels différents ont été testés pour ce système

TABLEAU 4: HUMIDITES RELATIVES DE SOLUTIONS SATUREES DE SELS

Solutions saturées de sels	HR
Chlorure de Lithium (LiCl)	11%
Chlorure de Calcium (CaCl2)	33%
Bromure de Sodium (NaBr)	59%
Chlorate de Sodium (NaClO3)	75%
Nitrate de Kalium (KNO3)	93%

A titre d'exemple, une solution saturée de $NaClO_3$ fourni normalement une humidité de 75%. Avec notre dispositif, l'humidité à l'intérieur de la chambre de mesure se stabilise à 64%±1%. 3 heures sont nécessaires pour atteindre cet état stable. Cela signifie qu'en cas d'ouverture de l'enceinte (pour placer les échantillons par exemple), le même temps d'attente est nécessaire pour obtenir le niveau d'HR souhaité. Pour des raisons pratiques, la nécessité de trouver un système qui permet une stabilisation plus rapide d'HR s'est imposée. Ainsi, nous avons placé dans la partie supérieure de la chambre de mesure un ventilateur afin d'accélérer l'homogénéisation du taux d'HR et de réduire le temps de stabilisation à 20 minutes.

L'avantage de ce système est qu'il ne nécessite pas d'infrastructures particulières. L'inconvénient majeur de cette configuration avec le ventilateur, provient du fait qu'on produit un flux d'air tourbillonnant. Ceci peut induire des variabilités difficilement contrôlables sur les propriétés de prise ou de perte en eau par le SC. De plus, le changement du taux d'HR se résume à quelques valeurs bien définies (tension de vapeur saturantes des solutions de sels) et implique le démontage du système pour changer la cloche. Ce processus est lourd et fastidieux.

Au vu de tous les points précités, il a été nécessaire de réorienter la conception de notre chambre à humidité contrôlée. Nous avons essayé de construire cette enceinte à HR contrôlée en se basant sur le principe de fonctionnement des chambres environnementales issu de la littérature : apport d'air sec et d'air humide [157, 158].

III.1.1.2 CONTROLEUR D'HUMIDITE RELATIVE 02 (RHC-02)

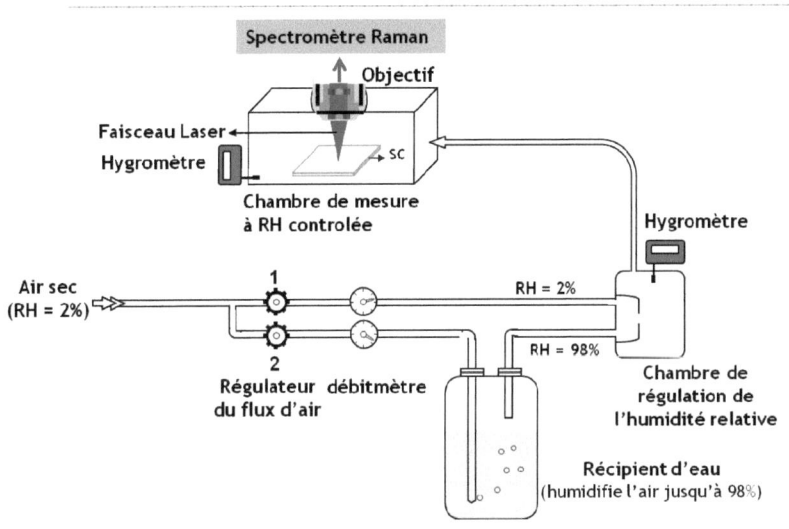

FIGURE 17: CONTROLEUR D'HUMIDITE RELATIVE 02 (RHC_02)

PRINCIPE DE FONCTIONNEMENT

Le fonctionnement du RHC_02 est basé sur un principe très simple de mélange d'air sec et d'air humide (Figure 17).

Le circuit gazeux (air sec à l'entrée HR= 2,5%) se divise en deux parties :

- Un circuit d'air sec muni d'une vanne 1
- Un circuit d'air également muni d'une vanne 2 fait passer l'air sec dans une bouteille contenant de l'eau (saturateur) ; cet air se charge d'humidité. Ce phénomène se traduit par la formation des bulles dans l'eau. L'air, saturé en humidité se trouvant au dessus de l'eau (humide à 98%), est acheminé par un autre tuyau.

Ces deux circuits aboutissent à une chambre de mélange équipée d'un thermo-hygromètre, elle-même reliée à une chambre sous le microscope du spectromètre Raman. Dans cette chambre, qu'on appellera chambre de mesures, est placé un thermo-hygromètre témoignant ainsi de la valeur de la température et d'HR pendant les acquisitions spectrales. Le taux d'HR (%) en sortie est ajusté en régulant les débits d'air sec et humide au niveau des vannes 1 et 2.

On remarque qu'un flux modéré dans la bouteille d'eau fournit de petites bulles et par conséquent une valeur élevée d'HR par rapport aux grosses bulles. Pour ce, nous avons opté pour un système fermé avec pores de petit diamètre afin de produire des bulles fines.

La bouteille contenant l'eau n'étant pas chauffée (contrairement à ce qu'on trouve dans la littérature), la température du système et de la chambre de mesure est indentique à celle de la salle climatisée: T=20°C±1°C.

TEST DE LA CAPACITE DU SYSTEME A CONTROLER L'HUMIDITE

Les tests ont été réalisés dans le laboratoire de chimie analytique à Chatenay-Malabry avec accès à l'air sec fourni par les compresseurs d'HR= 3,7% ; t° air sec = 23,8°C et la température de la pièce ≈ 23°C

Les différents taux d'HR ont été obtenus par réglages des débits d'air humide et d'air sec. Les mesures ont été notées au bout de 3 minutes de stabilisation à une valeur d'HR donnée.

TABLEAU 5: REGLAGE DE HR SUR LE SYSTEME RHC_02

Réglages des débits d'air		Chambre de mesures	T°	Evolution de la T°	
Air humide (%)	**Air sec (%)**				
100	0	89,2	23,3		
80	20	73,6	23,2		
65	35	58,7	23,6		
55	45	49,6	23,6		
45	55	40,3	23,6		
40	60	37,6	23,4	**moyenne**	**23,4**
35	65	32,8	23,5		
30	70	29,4	23,4	**CV**	**0,6%**
25	75	26,3	23,6		
20	80	20,8	23,5		
15	85	15,2	23,4		
10	90	9,5	23,2		
5	95	7,2	23,2		
0	100	3,7	23,8		

Les températures dans les boîtes restent voisines à celle de la pièce dans laquelle l'expérience est réalisée. Cela implique que, dans un premier temps, il suffit de travailler dans une salle climatisée pour stabiliser la température dans la chambre de mesure.

En conclusion, toutes les valeurs d'humidité comprises entre 3.7% et 90% peuvent être obtenues en utilisant le système développé ci-dessus (Tableau 5). Toutefois, en connectant directement le saturateur d'humidité à la chambre de mesure, on parvient à obtenir une HR de 98%. Ce système constitue le dernier prototype qui a été utilisé dans tous les travaux de cette thèse.

III.2.   OPTIMISATION DES PARAMETRES D'ACQUISITIONS

La micro spectroscopie Raman offre différentes possibilités d'analyse de l'échantillon, partant d'une analyse globale jusqu'à une analyse micrométrique ($1\mu m^3$ de volume). Il est également possible d'obtenir des informations aussi bien de la surface que celles des couches profondes avec des tranchages optiques grâce au système confocal. Ainsi, il est nécessaire d'optimiser les paramètres d'acquisition et de configurer l'appareillage en fonction des échantillons et du but poursuivi.

En plus de la résolution spatiale recherchée, le paramètre déterminant pour les analyses Raman reste le rapport "temps d'analyse / qualité du signal". On peut donc jouer sur le type de l'objectif de microscope, de l'ouverture du trou confocal et du réseau dispersif.

Nous avons procédé à l'optimisation des paramètres d'acquisition pour les échantillons de SC selon les critères suivants :

a) Choix de l'objectif du microscope adapté pour l'acquisition sur SC

b) Détermination du meilleur rapport résolution /sensibilité sur SC

c) Optimisation du temps d'acquisition et du binning (addition des signaux de plusieurs pixels).

a) Détermination de l'objectif adapté pour les acquisitions des spectres Raman sur le SC :

Le type d'objectif utilisé compte parmi les paramètres de mesure importants à optimiser. Celui ci est déterminant en termes de résolution spatiale, de distance de travail, et de qualité du signal. Dans cette partie, l'effet de l'objectif sur la qualité du signal ainsi la variabilité du signal Raman a été étudiée.

b) Optimisation de l'ouverture du trou confocal :

Ensemble avec l'objectif, l'ouverture du trou confocal détermine les dimensions du volume analysé. En diminuant son diamètre, le volume analysé est réduit, cependant le nombre de photons récoltés par le détecteur est moindre, d'où la diminution de l'intensité du signal. Donc, un bon compromis est nécessaire pour réaliser des mesures exploitables. En modifiant l'objectif et l'ouverture du trou confocal nous avons déterminé les résolutions spatiales.

C) Amelioration du temps d'acquisition et fusion des pixels (binning) :

Le temps a toujours un coût. Il est donc important de diminuer au maximum le temps nécessaire à la réalisation d'une tâche sans affecter la qualité des résultats. En vu de réduire d'avantage le temps d'acquisition, nous avons testé l'effet du « binning » du CCD de façon à diminuer le temps d'acquisition tout en gardant un bon rapport signal sur bruit. Le binning consiste à additionner les pixels (deux à deux ou plus) pendant la lecture du CCD. Aux temps d'acquisition égaux, cette addition de pixels conduit à une augmentation de l'intensité du signal. Par conséquent, il est possible de diminuer le temps d'acquisition. Toutefois, le binning résulte en une perte de résolution spectrale, ce qui n'est pas très gênant pour les bandes larges comme celle des élongations « OH ».

III.2.1   MICROSPECTROMÈTRE LABRAM HR

Dans un premier temps les mises au point spectrales ont été faites sur un microspectromètre Labram HR (Horiba Jobin Yvon), doté d'une source excitatrice de 633 nm ou de 785 nm ; d'un détecteur CCD (1024x256 pixels), et d'un réseau dispersif de 1800 traits/mm et de 4 types différents d'objectifs de microscope : 10X CF, 50X CF, 50X LF, 100X CF.

Dans les tests d'optimisation, les mesures ont été collectées approximativement au même point pour chaque échantillon avec trois répétitions par point. Nous avons fait varier un paramètre à la fois tout en maintenant les autres paramètres du spectromètre inchangés pendant l'expérience.

Nous avons constaté que **les rapports « signal/bruit »** de 50xCF et de 100xCF (objectifs à courte-focale) sont meilleurs par rapport au 50xLF (objectif à longue-focale). L'inconvénient des 2 premiers réside dans la difficulté de focaliser au risque même de toucher l'échantillon (à éviter pour l'étude de l'hydratation sur SC).

Parallèlement, la **meilleure résolution** est donnée par l'objectif 100x CF = 4 µm. Néanmoins, comme nous voulons observer une partie conséquente de l'échantillon, nous avons opté pour l'utilisation du **50xLF** avec un trou confocal **ouvert=1000 µm**.

L'autre avantage de ce dispositif, est sa facilité de manipulation grâce à sa distance de travail de 4 mm.

Nous avons ensuite évalué le temps d'acquisition optimum en se basant sur la qualité du signal. Nos analyses ont montré, que le temps d'acquisition représentant le meilleur rapport qualité de signal/ temps d'acquisition est 20 secondes, correspondant en temps réel à 2 minutes pour la **gamme 2800-3700 cm^{-1}**.

Pour réduire d'avantage le temps de mesure, nous avons eu recours au binning du CCD. Le Binning 16 associé à 2 secondes de temps d'acquisition a été choisi comme meilleure compromis : intensité du signal/temps d'acquisition. Néanmoins, la fusion d'un grand nombre de pixels affecte les pics fins et déforme l'information (pic de la phénylalanine à 1004 cm^{-1}). Le binning de 4 pixels seulement présente donc le meilleur compromis résolution/sensibilité/temps d'acquisition. Pour un réseau de 1800 traits/mm, un binning de 4 est acceptable pour obtenir des informations spectrales exploitables. Les analyses seront donc réalisées avec un binning 4 x 8sec.

En combinant tous ces résultats de mise au point, nous avons retenu les paramètres du Tableau 6 pour l'étude de l'hydratation du SC avec le spectromètre LabRam HR:

TABLEAU 6: PARAMETRES D'ACQUISITION SPECTRALE, LABRAM HR

Fente :	200 µm	Trou confocal :	1000 µm	Laser	633 nm
Réseau dispersif :	1800 traits	Objectif :	50xLF	Gamme :	2800-3700 cm^{-1}
Détecteur CCD	1024x256 pixels	Binning :	4	Temps :	8 S

Le spectromètre LabRam HR étant très résolutif, il génère un bruit important, nous essayons de le compenser par le binning avec toutes les conséquences que cela peut engendrer. De plus, avec l'usage du laser de 633 nm nous remarquons une fluorescence parasite non négligeable dans la partie des empreintes spectrales. En utilisant le laser de

785 nm, la fluorescence diminue mais la réponse du détecteur au-delà de 2600 cm⁻¹ devient très faible à cause de la réponse du CCD.

Ceci nous a conduit à tester d'autres systèmes instrumentaux afin d'obtenir un signal moins bruité, en conservant un temps d'acquisition réduit.

III.2.2 MICROSPECTROMÈTRE LABRAM

Cet appareil est doté d'un laser de 660nm. À cette longueur d'onde, la fluorescence parasite du spectre Raman est faible. De surcroît, de part son énergie faible, ce laser offre l'avantage de ne pas chauffer ou brûler l'échantillon. Il permet également d'avoir une réponse du détecteur correcte pour toute la gamme spectrale (400-3800 cm⁻¹).

Cet outil parait adapté à l'analyse de l'hydratation du SC (gamme 2600-3800 cm⁻¹) et à l'évaluation des effets de l'hydratation sur la conformation et l'organisation supramoléculaire des composantes du SC (gamme 400-1800 cm⁻¹).

Les mêmes tests d'optimisation identiques à ceux effectués pour le spectromètre LabRam HR ont été réalisés sur ce système, et les paramètres retenus sur cet appareil sont les suivants :

TABLEAU 7: PARAMETRES D'ACQUISITION SPECTRALE, LABRAM

Fente :	100 µm	Trou confocal :	200 µm	Laser	660 nm
Réseau dispersif :	950 traits/mm	Objectif :	100xLF	Gamme :	400-3800 cm⁻¹
Détecteur CCD	1024x256 pixels	Binning :	1	Temps :	20 s

COMPARAISION DES DEUX APPAREILLAGES

Nous avons comparé les résultats obtenus avec les deux appareils (LabRam HR à 633 nm et LabRam à 660 nm). On remarque que le signal obtenu à 660 nm comporte moins de fluorescence avec un rapport signal sur bruit supérieur et par conséquent plus facile à étudier (Figure 20), en particulier pour la zone correspondant aux empreintes spectrales (en dessous de 1800 cm⁻¹). Toutefois, les deux instruments permettent d'évaluer la quantité de l'eau dans le SC.

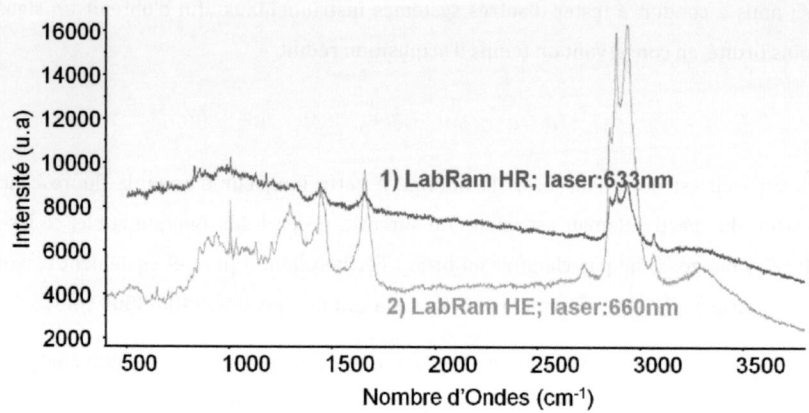

FIGURE 18: SPECTRES RAMAN DU *STRATUM CORNEUM*. 1) SPECTRE OBTENU AVEC UNE SOURCE LASER 633 NM, UN RESEAU DE 1800 TRAITS/MM. 2) SPECTRE OBTENU AVEC UNE SOURCE LASER DE 660NM, UN RESEAU DE 950 TRAITS/MM

RECAPITULATIF INSTRUMENTATION

Excitatrice	633 nm	660 nm
Réseau dispersif	1800 trais/mm	950 traits/mm
Binning	4	1
Temps d'acquisition	3 min	1 min
Gamme spectrale	400-3800 cm⁻¹	400-3800 cm⁻¹
Objectif	50XLF	100xLF

III.3. ÉTUDE DE LA VARIABILITE DU SIGNAL SPECTRAL RAMAN

OBJECTIFS :

Les objectifs de cette étude étaient de:

➢ Vérifier si les paramètres comme la localisation (point d'acquisition) et le temps d'irradiation par le laser ont un effet significatif sur les spectres Raman du SC.

➢ Déterminer le seuil de significativité au-delà duquel on pourra considérer que la différence n'est pas aléatoire ; qu'elle est due au facteur étudié (en l'occurrence l'humidité).

PROTOCOLE DE MANIPULATION:

Cette étude a été réalisée sur des échantillons du SC humain isolé par digestion enzymatique. Les spectres Raman ont été acquis à la surface de l'échantillon.

- Le SC a été placé sous microscope dans la chambre à HR contrôlée.
- Les spectres ont été acquis à 3 points de l'échantillon.
- A chaque point, 6 spectres ont été acquis à t_0, t_{5min}, t_{10min}, t_{30min}, t_{35min}, t_{40min}

CONDITIONS D'ANALYSE

HR=51.3±1%   température= 19.7±0.1°C

Fente :	200 µm	Trou confocal :	200 µm	Laser	633 nm
Réseau dispersif :	1800 traits	Objectif :	50xLF	Gamme :	200-3800 cm^{-1}
Détecteur CCD	1024x256 pixels	Binning :	4	Temps :	8 s

Après prétraitement des spectres, les rapports OH/CH ont été comparés comme décrit dans le Tableau 8 ci-dessous ; d'une part en fonction du temps (lignes), d'autre part en fonction des points (colonnes).

- Les points sont désignés par : p1, p2, p3
- Le temps est désigné par : t_0, t_5, t_{10}, t_{30}, t_{35}, t_{40}

TABLEAU 8: ETUDE DE LA VARIATION DU SIGNAL EN FONCTION DU TEMPS ET DU POINT ANALYSE. ANALYSE ANOVA

	p1	p2	p3	mean	S^2	CV en %			
t0	$2\ 511.10^3$	$2\ 184.10^3$	$2\ 380.10^3$	$2\ 359.10^3$	$27\ 145\ .10^6$	6.99	$\sum S^2 =$	$77\ 384.10^6$	
t05	$2\ 137.10^3$	$2\ 065.10^3$	$2\ 346.10^3$	$2\ 183.10^3$	$21\ 248.10^6$	6.68	$G_{obs}=$	0.350	$<G_{lim}=0.616$
t10	$2\ 126.10^3$	$2\ 144.10^3$	$2\ 213.10^3$	$2\ 161.10^3$	$2\ 124.10^6$	2.13	$S^2intra=$	$12\ 897.10^6$	
t30	$2\ 196.10^3$	$1\ 992.10^3$	$2\ 129.10^3$	$2\ 106.10^3$	$10\ 892.10^6$	4.96	$S^2IG=$	$37\ 337.10^6$	
t35	$2\ 149.10^3$	$1\ 971.10^3$	$2\ 058.10^3$	$2\ 059.10^3$	$7\ 961.10^6$	4.33	$F_{ANOVA}=$	2.8949	$<F_{lim}=3.11$
t40	$2\ 127.10^3$	$1\ 962.10^3$	$2\ 102.10^3$	$2\ 063.10^3$	$8\ 015.10^6$	4.34			
Mean	$2\ 208.10^3$	$2\ 053.10^3$	$2\ 205.10^3$						
S^2	$22\ 791.10^6$	$8\ 931.10^6$	$17\ 714.10^6$						
CV en %	6.84	4.60	6.04						
	$\sum S^2 =$	$49\ 437.10^6$							
	$G_{obs}=$	0.461	$<G_{lim}=0.707$						
	$S^2intra=$	$16\ 479.10^6$							
	$S^2IG=$	$47\ 135.10^6$							
	$F_{ANOVA}=$	2.8602	$<F_{lim}=3.68$						

Bilan :

L'analyse de la variance (ANOVA) nous montre que les moyennes ne sont pas significativement différentes que ce soit en comparant les points ou les temps d'acquisition. Ceci montre qu'il n'y a pas d'influence statistiquement significative du laser ou du point analysé.

CHAP IV. CARACTERISATION DE L'HYDRATATION DU *STRATUM CORNEUM*

IV.1.   PREAMBULE

Extérieurement une peau sèche se caractérisera par une structure écailleuse avec des microreliefs irréguliers ce qui se traduit par un aspect esthétiquement désagréable. Cependant malgré les symptômes cliniques clairement établis, les modifications au niveau moléculaire liées au taux d'hydratation du *SC* ne sont pas encore élucidées.

IV.1.1   QUANTIFICATION DE L'EAU DU *STRATUM CORNEUM*

Différentes techniques ont été jusqu'à présent utilisées pour déterminer le taux d'hydratation du SC: la Calorimétrie différentielle (DSC), la RMN du proton, la conductance, la capacitance, la microspectroscopie Raman etc,...

En 1986 Takenouchi et al ont utilisé la DSC pour étudier l'hydration du SC atteint de différentes pathologies comme : la xérose et l'ichtyose[159]. Ils ont démontré qu'il existe trois types d'eau dans le SC :

- **l'eau liée** : répartie en **eau fortement liée** (eau liée primaire restant dans le SC même
        à 0% d'HR) et **eau partiellement liée** (eau liée secondaire).
- **l'eau libre** sous forme de pentamères organisés en tétraèdres

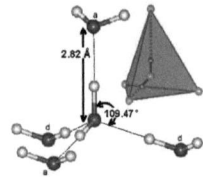

FIGURE 19: MOLECULES D'EAU SOUS FORME TETRAEDRIQUE

Dans le SC l'eau liée primaire représente 5mg/100mg de matière sèche (il faut chauffer pour l'extraire), tandis que l'eau liée secondaire pour un SC sain représente autour de 37% du poids du SC. Ces auteurs affirment qu'en dessous de 10%, le SC est rigide.

Néanmoins la DSC ne peut être utilisée que sur des échantillons isolés, limitant ainsi ses applications. En ce sens, la microspectroscopie Raman possède l'avantage de pouvoir :

- réaliser des analyses *in vivo*,
- analyser l'échantillon à différentes profondeurs,

- mesurer directement l'eau, et déterminer la conformation et l'organisation spatiale des molécules constitutives du SC (lipides, kératine).

En 1998 Gniadecka et al. ont utilisé le rapport (intensité O-H à 3250 cm⁻¹/ intensité C-H à 2940cm⁻¹) pour quantifier l'eau présente dans le SC par spectroscopie Raman [160]. En 2001 Caspers et al ont proposé le rapport d'une bande qui serait spécifique de la vibration d'élongation νOH (3350-3550cm⁻¹) et d'une bande spécifique de la vibration d'élongation νC-H des protéines (2910-2965 cm⁻¹) afin de mesurer l'hydratation cutanée[50]. D'autres auteurs comme Egawa ont repris ces bandes pour évaluer la quantité d'eau dans le SC. Néanmoins, aucun des rapports cités ci-dessus ne permet de distinguer l'eau libre de l'eau liée.

Il est bien connu que quand les liaisons hydrogènes sont fortes (eau liée), les vibrations d'élongation OH se déplacent de 3600cm⁻¹ vers 3300cm⁻¹ et si l'environnement des OH est bien défini, les pics seront fins et inversement [161]. Avec l'augmentation du taux d'HR, l'aire sous la courbe de la bande 3100-3700 cm⁻¹ du SC augmente en laissant apparaittre de nouvelles bandes sousjacentes.

Dans cette étude nous avons réalisé une analyse approfondie de cette région spectrale 3100-3700 cm⁻¹, par déconvolution de bande, afin de déterminer les vibrations spécifiques à l'eau non-liée et à l'eau liée aux autres composants du SC.

Pour ce, nous avons réalisé la déconvolution sur base de 4 sous bandes suivie d'une analyse des variations des aires des bandes déconvoluées en fonction de l'HR. La bande 3210 cm⁻¹ ne varie pas en fonction de l'HR alors que les bandes centrées sur 3285 cm⁻¹, 3345 cm⁻¹ et sur 3460 cm⁻¹ augmentent avec l'HR. Cependant, contrairement aux deux bandes du milieu, la surface de la bande à 3460 cm⁻¹ a une valeur nulle à HR inférieure à 10%.

En se basant sur des principes théoriques et sur des données bibliographiques démontrant la présence d'eau liée dans le SC même aux HR très faibles, sans chauffer [159] ; nous en déduisons que la bande à 3460 cm⁻¹ correspond à la vibration de l'eau non-liée. Les deux bandes autour de 3300 cm⁻¹ correspondent à l'eau en interaction avec d'autres entités moléculaires du SC (eau partiellement liée) ; Néanmoins, dans cette zone, nous pourrons avoir également la contribution des vibrations d'élongations NH à 3329 cm⁻¹. Finalement, la bande à 3210 cm⁻¹ qui ne varie pas en fonction de l'HR correspond à l'eau fortement liée (eau constitutionnelle). Ceci est en parfait accord avec la théorie indiquant que les OH liés vibrent à des nombres d'ondes inférieurs par rapport aux OH libres [161].

Sur la base de ces déconvolutions, nous avons déterminé, sur les spectres Raman du SC, les zones spectrales correspondant aux vibrations d'eau fortement liée, d'eau partiellement liée et d'eau non-liée.

- Eau fortement liée   : 3150-3245 cm^{-1}

- Eau partiellement liée : 3245-3420 cm^{-1}

- Eau non-liée   : 3420-3600 cm^{-1}

Les tests réalisés montrent que l'évolution en fonction de l'HR des aires des bandes déconvoluées et celle des zones sur spectres «bruts» (non -déconvolué), donnent des résultats comparables. **Ceci permet de valider ces zones pour la quantification d'eau liée ou d'eau non-liée du SC directement sur base d'un spectre Raman (non -déconvolué).**

IV.1.2   L'EFFET DE L'HUMIDITE RELATIVE SUR LA STRUCTURE DES LIPIDES ET DES PROTEINES DU *STRATUM CORNEUM*

Il est bien connu que l'élasticité et l'efficacité de la fonction barrière du SC sont fortement affectées par son degré d'hydratation [162, 163]. Par conséquent, il est d'un grand intérêt d'examiner l'interaction de l'eau avec les constituants du SC. Les molécules d'eau interagissent avec les groupements polaires des molécules du SC augmentant par la même occasion les espaces intermoléculaires et intramoléculaire (structure secondaire des protéines)[50].

Certains auteurs affirment qu'il n'y a pas d'incorporation de l'eau dans les lipides [164, 165]. D'autres ont démontré qu'un contact prolongé conduit à l'apparition d'eau libre dans la matrice lipidique [166, 167]. L'interaction eau-lipides, probablement au niveau des têtes polaires, pourrait conduire à une modification de la conformation de chaînes hydrocarbonées et surtout de leur organisation supramoléculaire.

Les conformations des lipides du SC sont souvent évaluées, en spectroscopie Raman, par la comparaison du pic à 1080 cm^{-1} (témoin des conformations *gauches*) et les pics centrés sur 1060 et 1130 cm⁻¹ (témoin des conformations *trans*). Contrairement aux *gauches*, les *trans* sont associés à un état plus compact.

L'organisation supramoléculaire des lipides du SC est classiquement évaluée en spectroscopie Raman par les vibrations de cisaillement autour de 1445 cm⁻¹. Cependant, la contribution du signal de la kératine dans cette zone spectrale rend difficile son exploitation pour décrire l'organisation des lipides. Certains auteurs attribuent les vibrations d'élongation des CH_2 (entre 2800 et 3000 cm⁻¹) à l'organisation [160].

La structure spatiale d'une protéine détermine ses propriétés physico-chimiques comme l'élasticité ; ainsi que ses fonctions.

La conformation de la kératine est fréquemment identifiée sur la base de ses propriétés en diffraction aux rayons X : hélices α, feuillets ß, des structures aléatoires non ordonnées. Les différents types de conformations pour les fonctions amide I, II et III ont leurs propres caractéristiques vibrationnelles. En général, ces structures sont formées par les liaisons hydrogènes intramoléculaires entre l'oxygène du groupement carbonyle au niveau de la liaison peptidique et l'atome hydrogène d'un autre.

Ces deux structures organisées peuvent être affectées par l'addition de molécules d'eau. Les liaisons hydrogènes peuvent être affaiblies par la présence de molécules d'eau s'insérant elles-mêmes entre les chaînes peptidiques. D'autres groupements polaires de la protéine établissent également des liaisons hydrogènes avec l'eau.

ARTILE 1 : EFFETS DE L'HUMIDITE ATMOSPHERIQUE SUR LA STRUCTURE MOLECULAIRE DU *STRATUM CORNEUM* : ETUDE *EX VIVO* PAR MICROSPECTROSCOPIE RAMAN

Contexte : La sécheresse cutanée se caractérise par diverses manifestations cliniques. Toutefois, son effet au niveau moléculaire n'est pas encore complètement élucidé. Au cours des dernières décennies, différentes méthodes ont été employées pour étudier l'hydratation de la peau [1, 4, 46, 166, 168, 169]. En plus des techniques «classiques» comme la perte insensible en eau (PIE), la conductance de la peau, profilométrie de la peau et l'analyse de desquamation, des méthodes moins «classiques» ont également été utilisées (spectrocopy Raman, RMN etc...). Il est connu que l'eau existe dans le SC sous trois types: eau fortement liée, eau partiellement liée secondaire et eau non liée aux autres types de molécules [159].

Méthodes : Dans ce travail, des mesures *ex vivo* ont été effectuées à différents niveaux d'HR. La teneur globale en eau dans le SC ainsi que l'eau liée et l'eau non liée ont été mesurées séparément en utilisant les bandes des vibrations d'élongation OH (vOH) situées entre 3100 et 3700 cm^{-1}. Les conformations de lipides du SC ont été évaluées à l'aide de la position maximale de torsion CH$_2$ (δCH$_2$) (1296 cm^{-1}) [67, 170, 171] et en comparant les pics à 1060 et 1130 cm^{-1} (élongations C-C : vC-C de conformations trans) associés à l'état compact, et le pic à 1080 cm^{-1} (vC-C de conformations gauches) qui représente les formes désordonnées des lipides. L'organisation latérale des lipides du SC a été étudiée en utilisant les bandes d'élongation CH (vCH) dans l'intervalle 2800-3000 cm^{-1}.

Pour analyser l'effet de l'humidité sur les conformations de protéines, nous nous sommes intéressés à la région Amide I (1600-1700 cm^{-1}) et dans le doublet de Fermi de Tyrosine (850/830 cm^{-1}) ainsi que dans les vibrations v_{asym}CH$_3$ (2932 cm^{-1}).

Résultats : Nos résultats ont montré que la quantité d'eau et sa structure au sein du SC varient systématiquement en fonction des HR avec une évolution différente des 3 types d'eau. La structure de la matrice lipidique aussi bien que celle des protéines évolue proportionnellement à l'eau partiellement liée. La meilleure organisation des lipides (conformation trans élevé, compacité importante) a été observée pour des niveaux d'HR intermédiaires autour de 60%.

Conclusion : Ce travail prouve que l'étude de l'eau liée et en particulier l'eau partiellement liée est d'un grand intérêt dans la compréhension de l'effet de l'hydratation sur la fonction barrière. En effet, les résultats montrent que la structure des protéines ainsi que la compacité de la barrière lipidique du SC sont directement liés à la variation de l'eau partiellement liée.

ARTICLE 1

Analyst

RSC Publishing

PAPER

Cite this: *Analyst*, 2013, **138**, 4103

Effects of atmospheric relative humidity on Stratum Corneum structure at the molecular level: *ex vivo* Raman spectroscopy analysis

Raoul Vyumvuhore,[a] Ali Tfayli,*[a] Hélène Duplan,[b] Alexandre Delalleau,[b] Michel Manfait[c] and Arlette Baillet-Guffroy[a]

[a]*Group of Analytical Chemistry of Paris-Sud (GCAPS), Faculty of Pharmacy, Univ. Paris-Sud, Chatenay-Malabry, France*

[b]*Center of Research Pierre Fabre Dermo-Cosmetics (PFDC), Toulouse, France*

[c]*MéDIAN Unit, CNRS UMR 6237, Faculty of Pharmacy, Univ. Reims Champagne Ardennes, Reims, France*

Acknowledgment: ANR-12-JSV5-0003 CARE; Pr Pierre CHAMINADE

Abstract:

Skin hydration plays an important role in the optimal physical properties and physiological function of the skin. Despite the advancements in the last decade, dry skin remains the most common characteristic of human skin disorders. Thus, it is important to understand the effect of hydration on *Stratum Corneum* (SC) components. In this respect, our interest consists in correlating the variations of unbound and bound water content in the SC with structural and organizational changes in lipids and proteins using a non invasive technique: Raman spectroscopy. Raman spectra were acquired on human SC at different relative humidity (RH) levels (4%-75%). The content of different types of water, bound and free, was measured using second derivative and curvefitting of the Raman bands in the range 3100-3700 cm^{-1}. Changes in lipidic order was evaluated using νC-C and νC-H. To analyze the effect of RH on the protein structure, we examined in the Amide I region, the Fermi doublet of Tyrosine, and the $\nu_{asym}CH_3$ vibration. The contributions of totally bound water were found not to vary with humidity, while partially bound water varied with three different rates. Unbound water increased greatly when all sites for bound water were saturated. Lipid organization as well as protein deployment was found to be optimal at intermediate RH values (around 60%) which correspond to the maximum of SC water binding capacity. This analysis highlights the relationship between bound water, the SC barrier state and the protein structure and elucidates the optimal conditions at ambient conditions. Moreover, our results showed that increased content of unbound water in the SC induces disorder in the structures of lipids and proteins.

Keywords: *stratum corneum* hydration; bound water; protein structure; lipid organization; Raman spectroscopy

Graphical abstract

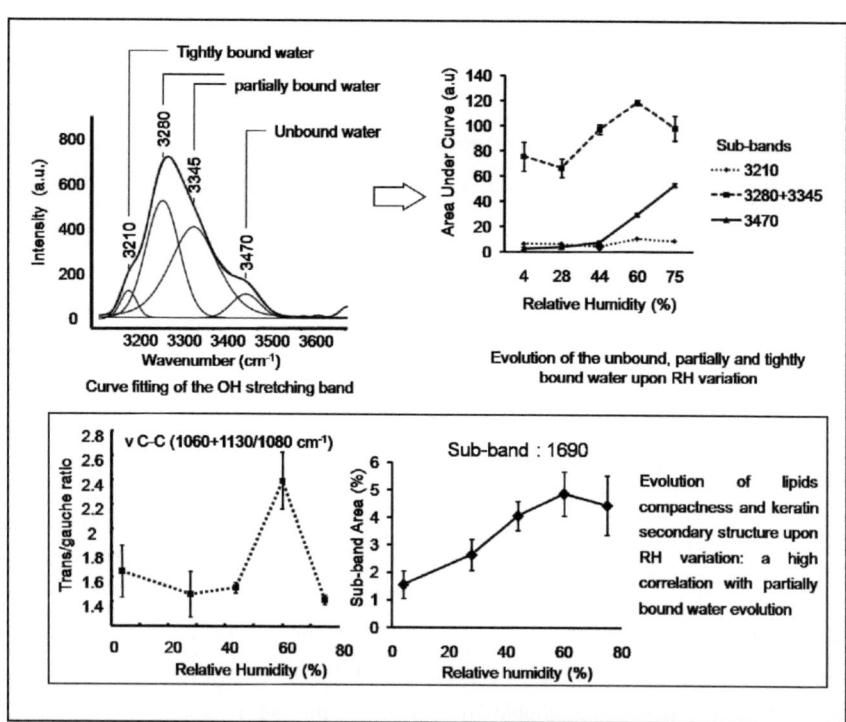

Introduction

Skin hydration has been widely studied over the last decades to improve the understanding of "dry skin" phenomenon usually met in dermatology and in cosmetic applications, particularly in pathologies like ichthyosis and xerosis. In 1994, Warner et al. identified the precise layer of the skin that loses water during the drying process, namely the outermost layer of the skin "*Stratum Corneum* (SC)" [1].

The SC is composed of three major components: corneocytes-anucleated cells filled with keratin filaments as well as natural moisturizing factors (NMF), corneodesmosomes-tight junctions between corneocytes and an intercellular lipid bilayer matrix. The SC controls the skin hydration by three different mechanisms: barrier function, water binding and water diffusion within the SC. The skin barrier maintains the level of water required for normal physiological function. This barrier function is mainly due to the lipids in the extracellular matrix of the SC. The lipid matrix include ceramides (CER), free fatty acids (FFA), and cholesterol (CHOL), which are present in nearly equimolar ratios [2]. The lateral packing of these lipids is thought to be important for their ability to maintain the barrier function, as well as for maintaining a desirable skin feel and plasticity [3]. A pure liquid crystal system or a solid system of SC lipids causes rapid water loss through the bilayer. Maintaining the balance between the two phases is required for optimal barrier function in preventing water loss [4]. Moreover, the presence of NMF, which are hygroscopic molecules, enables the absorption of water and maintains the acidity at the surface of the SC that is essential for the regulation of the activity of lipase enzyme releasing SC lipid components. Finally, the maturity of corneodesmosomes influences the tortuosity of the SC and consequently water distribution.

Despite the advancement in the understanding of the structure, composition and function of the SC, dry skin remains the most common symptom of human skin disorders. Thus, it is important to understand the effect of hydration on SC components at the molecular level.

Over the last few decades, different methods have been employed to study skin hydration [5-10]. In addition to the "conventional" techniques as transepidermal water loss (TEWL), skin conductance [9], skin profilometry and desquamation analysis, less "conventional" methods were also investigated. Using Differential Scanning Calorimetry (DSC), Takenouchi et al. demonstrated the presence of three types of water in the SC: primary bound water, secondary bound water and bulk water [11]. In addition to that, It has been shown that bound water quantity determines the flexibility and the stiffness of the SC. Below 10 % (g water g^{-1} tissue) , the SC is stiff [12, 13]. For normal SC, this quantity would be around 33 % (g water g^{-1} tissue) while a diseased SC contains less bound water [11, 14, 15].

In addition to water type identification and quantification, the effect of water content on SC composition and physiology has been investigated [16, 17]. Many studies have illustrated that hydration induces corneocytes swelling [18-20]. Since water is an elasticity enhancer for the SC, water-keratin interactions has been investigated using techniques as NMR [21, 22] and cryo-scanning electron microscopy [20, 23].

Interactions between water and SC lipids have also been studied using cryo-scanning electron microscopy. These works revealed that, at high SC hydration levels, water was located inside the corneocytes and in their intercellular regions [1, 23, 24]. Other studies suggested that the water-lipid interaction probably occurs with the polar heads of lipids. Thus, it could lead to a modification of hydrocarbon chains conformation and especially could change their supramolecular organization [10, 25-27]. A correlation between SC lipid structure and water flux through the SC was also suggested by Golden et al. based on changes in infrared spectra [28].

One of the main advancements in understanding skin hydration was the introduction of Raman spectroscopy. Raman spectroscopy is a non-invasive vibrational technique which provides direct information on structure and organization of the SC components [29-32]. The main advantage of this technique is that the same information can be obtained *ex vivo* and *in vivo*. Developments performed by the team of Puppels [33, 34] enabled direct evaluation of skin hydration by using the ratio of Raman intensities at 3390 cm^{-1} (water) and 2935 cm^{-1} (proteins). This ratio was a key point for different further studies to determine, among others, the water gradient, the SC boundary and thickness as well as the effects of different moisturizers on skin hydration [31, 33, 35]. In parallel, Gniadecka et al. have attempted a separate quantification of free and bound water but they were limited by the spectral range of their measurement [31, 36]. In addition to information on water and NMF contents [37], the potential of Raman spectroscopy to determine the conformational order and the lateral packing of lipids [30, 38-43] and the secondary structure of Protein [44-46] has been widely described.

To the extent to our knowledge, no Raman analysis has been performed to date on the SC to study the variations of water structures, i.e. unbound and bound water, as a result of systematic variations of atmospheric relative humidity (RH); and to correlate these variations with structural and organizational changes in lipids and proteins.

In the current study, Raman spectroscopy was used to determine the effect of the atmospheric RH on the SC hydration and on its relationship with the lipid conformational order and the protein structure. *Ex vivo* measurements were performed at different RH values. Global water content in the SC as well as bound and unbound water were measured separately using the OH stretching (νOH) band (3100-3700 cm^{-1}) [31, 47-49]. Conformations of SC lipids were evaluated using the maximum position of CH$_2$ twisting (δCH$_2$) (1296 cm^{-1}) [30, 38, 40] and by comparing the peaks at 1060 and 1130 cm^{-1} (C-C stretching (νC-C) of trans conformations) associated with ordered state and the peak at 1080 cm^{-1} (νC-C of *gauche* conformations) representing disordered forms of lipids [32, 38, 42]. The lateral packing of SC lipids was studied using the CH stretching (νCH) in the 2800-3000 cm^{-1} range [30, 31, 43, 50-52]. To analyze the effect of RH on protein conformations, we were interested in the Amide I region (1600-1700 cm^{-1}) [36, 44, 45, 53-55] and in the Fermi doublet of Tyrosine (850/830 cm^{-1}) as well as in ν_{asym}CH$_3$ vibrations (2932 cm^{-1}) [33, 36, 44]. Our results have shown that the amount of water and its structure within the SC was seen to vary systematically as a function of RH and that changes in the spectral features associated with the structure of lipids and proteins were seen to change in a correlated fashion.

Materials and methods

Stratum corneum isolation

Human abdominal skin samples were obtained after plastic surgery from 5 female patients aged between 40 and 50 years. Samples were stored at -20°C for less than 24 hours. SC samples were prepared using enzymatic digestion method [56-60]. The subcutaneous fat and connective tissues were first removed. The skin was then immersed in distilled water (Milli-Q reagent water system) at 60°C for 1 minute. The epidermis was subsequently removed from the dermis. The epidermis was then placed; SC sided up, onto a filter paper imbibed with a 0.2% trypsin solution (0.2% in distilled water; Sigma-T4665) for 1 hour. The SC was carefully separated from the underlying epidermis with tweezers. The sheet of SC was spread in a water container at ambient temperature and was settled onto a greaseproof paper. Then, the isolated SC was dried under vacuum in desiccators (containing P_2O_5). Three SC samples were prepared from each patient.

Humidity controlled environment

A humidity controlled chamber, based on dry air/ wet air mixture, was specifically designed to fit the microscope head. Air flow was controlled by volume flow controllers DYNAVAL AIR (AIR LIQUIDE, Paris, France). A hygrometer/thermometer Dewpoint Pro (Control Company, Friendswood, USA), with ± 2% of accuracy, was used to measure the relative humidity. The temperature of the air-conditioned room was maintained at 20°C±1°C.

For our study, RH was set to 4%, 28%, 44%, 60%, 75%. Before spectral analysis, the samples were allowed to equilibrate for 4 hours at each RH level.

Raman microspectrometer

Samples were placed in a humidity controlled chamber under the microscope interfaced to a confocal Raman microspectrometer LabRam (Horiba Scientific, Lille, France). A video image of the sample was used for accurate positioning of the laser spot on the sample. A 660 nm pumped Nd:YLF laser (TOPTICA PHOTONICS, Munich, Germany) giving a 10 mW power, on the sample, was used. The 660 nm excitation wavelength was chosen because it gives a weak fluorescence-like background in the fingerprint region [61] and a high Raman Stokes signal in the high wavenumbers region. A long focal microscope objective PL Fluotar L 100X/ NA 0.75 WD 4.7 (Leica, Mannheim, Germany) was used to focus the laser light on the surface of the sample and to collect the back scattered light. Confocal pinhole was set to 150 μm. The in-depth spot size was around 8 μm; the in-depth resolution was measured using the method described by bruneel et al. [62, 63]. The collected light was filtered through a notch filter and dispersed with a 4 cm^{-1} spectral resolution using a 100 μm slit and a holographic grating of 950 grooves/mm. The Raman Stokes signal was recorded with a Charge-Coupled Device detector: CCD camera (Andor technology, Belfast, UK) containing 1024 x 256 pixels. Spectral acquisition was performed using Labspec 5 software (Horiba Scientific, Lille, France). Raman measurements were performed on the SC surface in the 400–3800 cm^{-1} spectral range. For each scan, a 20 sec exposure time was used. 15 samples from 5 patients were measured, for each RH value, 6 measurements per sample were performed.

Raman signal processing

All spectra were subjected to the same automated preprocessing protocol. All spectra were smoothed using savitsky Golay algorithm on 11 points [64] and baseline corrected using an automatic polynomial function [65]. Instrument dependent spectral variations were corrected by a normalization process. To compare spectra from different RH levels, normalization must be done on a humidity independent band. Given that the protein structure may be influenced by RH variations, using a protein specific vC-H stretching feature for normalization may induce errors in the pre-processing. The same errors can occur using a unique lipid vC-H stretching band. However, since vC-H stretching bands (2800-3000cm^{-1}) are interrelated, a balance of the integrated intensity values of these peaks can be observed and the global area of the 2800-3000 cm^{-1} band can be considered as invariant with RH. The difference in signal intensity can be compensated by normalizing on this band.

Curve fitting

Curve fitting of the OH stretching band and Amide I (AI) band of the SC Raman spectra have been carried out using the Least Squares Fitting protocol of the Labspec 6 software (Horiba Scientific, Lille, France). The software allows the user to identify a number of sub-bands within a spectral region using the second derivative. It then automatically adjusts the combination of bands to best fit the spectral profile. Spectra were initially constructed with a combined Lorentzian and Gaussian function [45]. The maximum shift of the sub-band position and bandwidth was set to ±5 cm^{-1} and ±15 cm^{-1} respectively, while intensity was left free to adapt to the fit. The quality of the fit was estimated by the standard error and the χ^2 values.

For vOH sub-band quantification in the 3100-3700 cm^{-1} range, a prior normalization was done on vCH vibration band and the area of each sub-band was calculated, while for the determination of the secondary structure content using Amide I band, the area of each component was divided by the sum of the area of all Amide I components.

Statistical analysis:

The analysis of variance (ANOVA) is one of the most widely used tests for experimental data. It is a powerful method to compare differences between the means of several groups. The test compares the variability of each population (intra-group variances) and variability of grouped populations (inter-group variance) and thus determines whether the groups are significantly different. For each RH level, the average results of the measurements on all patients' SC were presented as mean ± Standard Deviation and an ANOVA test was performed.

Results and discussion

Water content quantification

Figure 1 presents the high wavenumber region (3100-3700 cm^{-1}) of the spectra of a SC exposed to different RH levels. This region contains water vOH bands with the contribution of OH and NH stretching vibrations issued from the SC proteins and lipids. For water quantification in SC, Caspers et al. [33] used the integrated intensities of **the band under 3350-3550 cm^{-1}**. As mentioned by Caspers et. al.; this region was previously used for water quantification in eye lenses and cornea [66, 67]. The band under 3350-3550 cm^{-1} is directly associated with vOH of unbound water present in the studied samples [31, 47]. The other spectral features were ignored in order to avoid the contribution of NH vibrations [68]. Meanwhile, as shown in figure 1, the evolution of the 3100-3700 cm^{-1} region is not limited to the 3350-3550 cm^{-1} part. The intensity of the 3100 -3350 cm^{-1} region, associated to vOH issued from bound water molecules, seems to be RH dependent. Therefore, it is worth studying separately the different spectral features in the 3100-3700 cm^{-1} region for a reliable analysis of both unbound and bound water using band deconvolution.

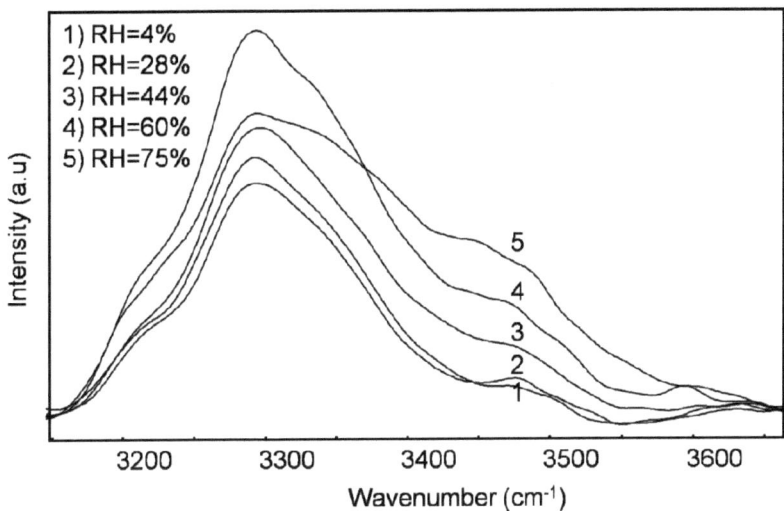

Figure 1: Mean *ex vivo* Raman spectra of one patient's SC in the 3100-3700 cm^{-1} spectral range at different relative humidity levels (4%, 28%, 44%, 60%, and 75%).

Deconvolution of OH band

The spectral features associated with the different types of water and the contributions of NH vibration were assessed by performing a band deconvolution. Four different bands were identified by the second derivative (3210 ±5 cm^{-1}, 3280 ±5 cm^{-1}, 3345 ±5 cm^{-1} and 3470 ±5 cm^{-1}). The obtained Gaussian/Lorentzian sub-bands are presented in figure 2.

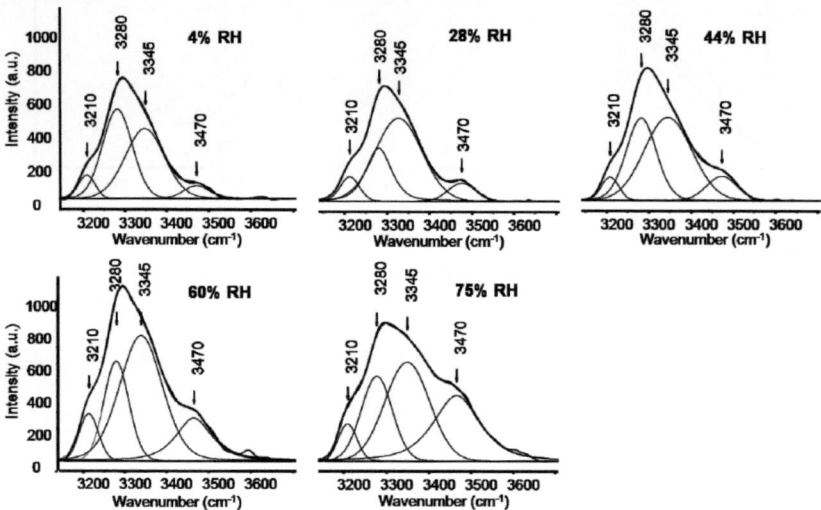

Figure 2: Curve fitting of the νOH band from one patient's SC mean spectra at different RH values. Four sub-bands were identified: 3210±5 cm^{-1}; 3280±5 cm^{-1}; 3345±5 cm^{-1} and 3470±5 cm^{-1}

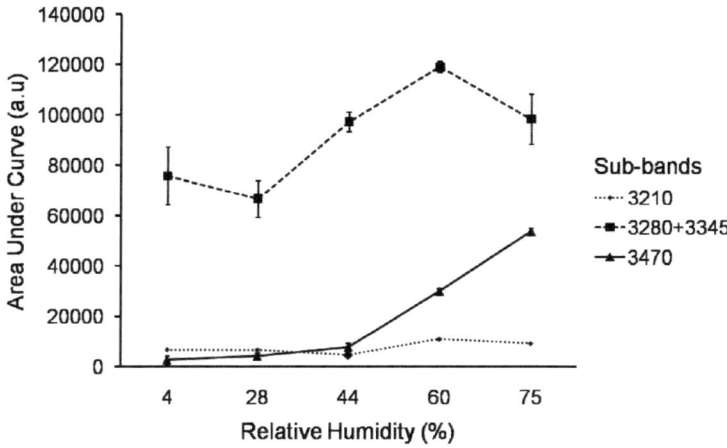

Figure 3: Evolution of the integrated area of vOH sub- bands along RH variations. The partially bound water associated sub-bands (3280 and 3345 cm⁻¹) were taken together.

Evolution of the 3210 cm⁻¹ sub-band: totally (primary) bound water.

In addition to NH stretching, the 3210 cm⁻¹ sub-band can be related to very tightly bounded water (primary bound water) as described by Qiang Sun et al. [48]. In this case, water molecules are involved in double acceptor and double donor hydrogen bonds giving an vOH band around 3220 cm⁻¹ [48]. The intensity of this sub-band does not vary along changes in RH (P-value= 0,894) (Figure 3).

Evolution of the 3280 cm⁻¹ and 3345 cm⁻¹ sub-bands: partially bound water.

Totally bound water is tightly bounded to the SC polar sites of proteins constituting a first monolayer of water, whereas partially bound water is bound to the first monolayer and to other SC molecular components [69]. This means that water molecules interact partially with neighboring molecules using only 2 or 3 of the 4 possible hydrogen bonds. The sub-bands at 3280 cm⁻¹ and 3345 cm⁻¹ are associated with partially bound water with different number and types of hydrogen bonds.

The evolution of the global partially bound water content was therefore observed by plotting the sum of the AUC of both sub-bands (Figure 3). Between 4% and 28% RH, the partially bound water content did not vary significantly. This was confirmed by an

ANOVA test with a P-value of 0.949. From 28% to 60%, it rose sharply (P-value=0.0000009) before decreasing above 60 % RH (P-value=0.00002). One can assume that at high levels of RH, unbound water weakens the forces of intermolecular bonds between water and SC components which results in a decrease of the features at 3280 cm^{-1} and 3345 cm^{-1}.

These observations are in agreement with Differential Scanning Calorimetry (DSC) literature data reporting that, above 28% RH, additional water layers are formed on top of the first monolayer [69]. They are also in accordance with NMR [21] and infrared spectroscopy [13] studies on the mechanism of SC water uptake.

Evolution of the 3470 cm^{-1} sub-band: Unbound water.

In the present work, unbound water is considered as water that is not directly linked to the SC components; thus presenting no hydrogen bonds with SC lipids or proteins [23, 70]. This can be defined as water present in a multilayer system (second layer, third layer...) which can be found at low RH values [15, 69]. OH stretching associated with this water arises around 3470 cm^{-1} [49]. From 4% to 28% RH, we observed a small appearance of the band around 3470 cm^{-1} which increased slightly up to 44% RH. Over 60% RH the integrated intensity of this band increased significantly and proportionally with relative humidity (Figure 3). This is in concordance with the results of Pieper et al. that have shown that the H$_2$O uptake at RH 60% corresponds to four layers of water [69]

Given the complexity of the molecular organization of keratin and lipids in the SC; water distribution in the SC is heterogeneous [71]. In other terms, regions with only one water layer (totally bound water) can coexist with region with two or more layers (partially bound and unbound water) [15].

Folded molecules present mainly totally bound water. Upon hydration, keratin uncoils and makes available new sites for partially water bounding and further. This could explain the simultaneous increase of partially bound water and unbound water (Figure 3) up to 60 % RH. At this RH, one can suggest that H-bonding sites of SC are saturated. For higher RH levels, unbound water content increases and loosens the H-bonds between partially bound water and SC components as previously mentioned.

Effect of RH on the organization of SC lipids and proteins

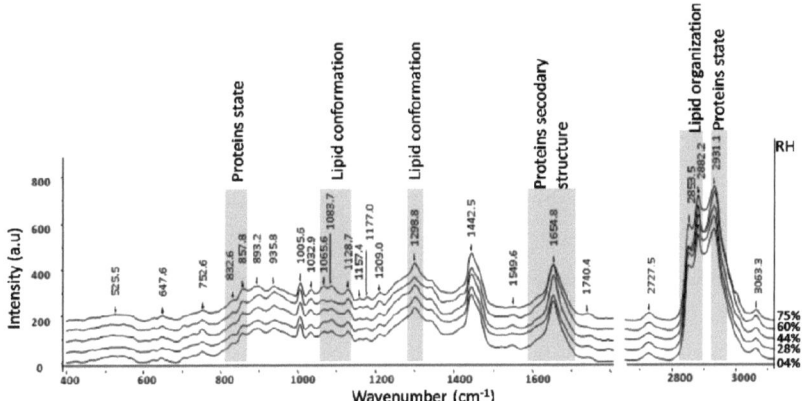

Figure 4: Mean *ex vivo* Raman spectra of human *stratum corneum* at different relative humidity levels (4%, 28%, 44%, 60%, and 75%) in the range of 400-3120 cm^{-1}.

Effect of RH on the organization of SC lipids

Three different spectral features were used in order to study the lipid conformational order in the SC (Figure 4):

1- νC–C modes between 1040 cm^{-1} and 1145 cm^{-1}: *trans/gauche* conformers ratio was observed by calculating $I_{1060+1130}/I_{1080}$ ratio (Figure 5A). High values of this ratio are associated with a compact state in the lipid packing while a decrease is indicative of a loosening [32, 38, 42].

2- δCH$_2$ at 1298 cm^{-1}: the position of this feature is sensitive to the conformational disorder (Figure 5C). A shift to higher wavenumbers is indicative of an increase of the disorder and higher *gauche* conformers content [30, 38, 40].

3- ν_{asym}CH$_2$ (2882 cm^{-1})/ν_{sym}CH$_2$ (2852 cm^{-1}) ratio: this ratio is generally used as an indicator of the conformational state and the lateral packing [30, 43] (Figure 5B). High values are associated with higher *trans* content and a compact organization.

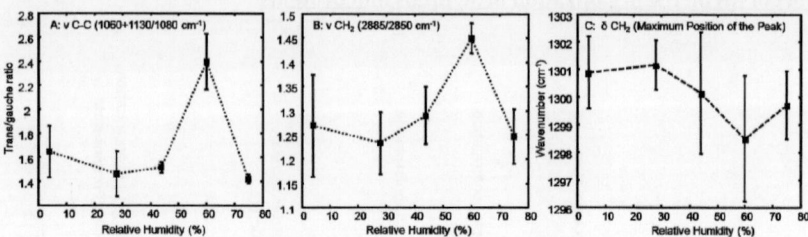

Figure 5: Conformation of *stratum corneum* lipids at different atmospheric RH values (3%, 28%, 44%, 60%, and 75%). Mean data at each RH level with n=6..The conformational changes were estimated using descriptors such: A) Band area of the νC-C *trans/gauche* ratio= $S_{1060+1130}$ / S_{1080}; B) The νCH$_2$ *trans/gauche* ratio= I_{2885} / I_{2850}; C) The maximum of δCH$_2$ twisting peak around 1298 cm^{-1}.

All three graphics shown in Figure 5 present a similar evolution in the *trans/gauche* ratio. From 28-60% RH, the *trans/gauche* ratio increases with the humidity and then falls down at high humidity levels (> 70 %). At 4%RH, lipid matrix is slightly more ordered than at 28% RH, suggesting that the skin is responding to the drying environment by offering a less entangled barrier. It is worth noticing that the evolution of *trans/gauche* ratio is correlated with the evolution of the partially bound water content (figure 3).

A possible explanation of the observed modifications can be proposed as follows: At intermediate RH values, the amount of partially bound water is suitable for an optimal organization of lipids. For high hydrated SC, unbound water molecules are inserted in the polar-head space of lipids decreasing the hydrogen bonds forces of the partially bound water and modifying spaces between hydrocarbon chains. This could lead to an increase in the fluidity (Figure 5) and confirms previous results obtained using neutron scattering [10], cryo-scanning electron microscopy [23], freeze-fraction electron microscopy [26, 27], NMR and sorption microcalorimetry [22, 69]. Such findings could be compared to those of Alonso et al. on mechanical tensions in the SC, suggesting that elasticity parameters change mainly for intermediate RH values and that only secondary bound water is responsible of the SC flexibility [12, 13]. The opposite occurs when SC undergoes little water content, the lipid molecules interact strongly and its hydrocarbon chains become entangled.

Effect of RH on proteins conformations

In order to highlight the effect of hydration on the protein structure, the evolution of the Fermi doublet of Tyrosine, the $\nu_{asym}CH_3$ band and the Amide I (AI) band has been studied.

Fermi doublet of Tyrosine and $\nu_{asym}CH_3$ vibrations

The doublet located at ~830 cm⁻¹ and ~850 cm⁻¹ is caused by the Fermi resonance between the ring-breathing vibration and the overtone of an out-of-plane ring-bending vibration of the Para-substituted benzenes. The relative intensity ratio of I_{850}/I_{830} is sensitive to the property of the hydrogen bonding of the phenolic hydroxyl group [44, 72] . High values of this ratio are indicative of exposed tyrosine while a shift for lower values is related to buried tyrosine (figure 6).

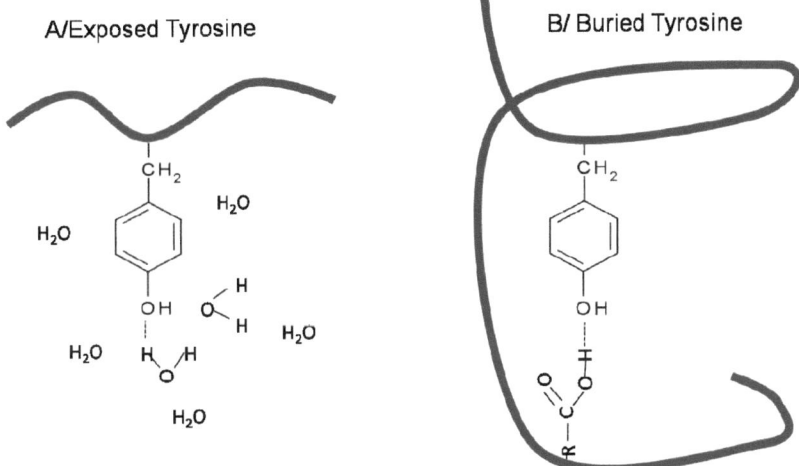

Figure 6: Two typical configurations of Tyrosine residues in a protein, namely the A/Exposed Tyrosine and B/Buried Tyrosine.

In addition to that, the shift of $\nu_{asym}CH_3$ towards high wavenumbers testifies to the unfolded status of the keratin so favored due to the interaction between the side chains of its amino acids and water molecules[31, 36].

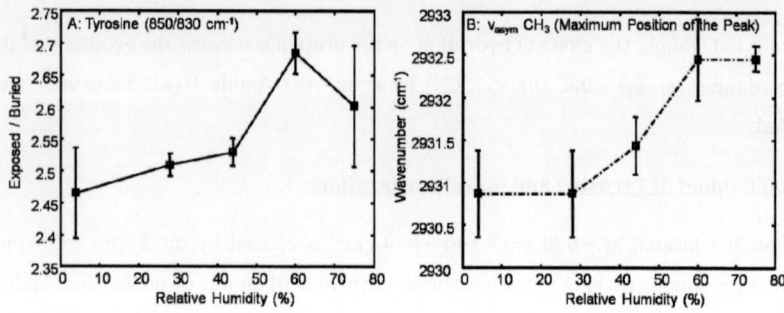

Figure 7 : A) Ratio I_{850} / I_{830} as a function of HR; B) Maximum position of the band centered around 2932 cm^{-1} as a function of HR; mean data at each RH level with n=6.

As can be observed in Figure 7A, the evolution of the intensity ratio 850 cm^{-1} / 830 cm^{-1} upon the dehydrating process can be divided in three regions. When the RH decreased from 75% to 60%, the bound water amount (in interaction with the keratin) did not vary (Figure 3). Interestingly, in that range the 850 cm^{-1} / 830 cm^{-1} ratio did not vary significantly suggesting that the mole fraction of Tyrosine that was interacting with water molecules remained stable. This is additionally supported by the fact that in the same conditions the indicator of the folding/unfolding process of protein (maximum position of $v_{asym}CH_3$: 2932 cm^{-1}) remained unchanged. Below 60% RH, the bound water content (Figure 3) as well as the amount of exposed Tyrosine decreased (Figure 7A), this is accompanied with a folding process of the protein (shift of $v_{asym}CH_3$ towards lower wavenumbers) (figure 7B). The loss of the bound water destabilizes the protein structure and conformation. When the bound water was removed the protein folded and then the Tyrosine appeared in a buried conformation (figure 8A). Studying structural changes on insulin in dehydration process, Zeng et al [44] have reported similar phenomena. This is in conjunction with scanning force microscopy and the cryo-scanning electron microscopy studies which demonstrated that the thickness of isolated corneocytes increases upon soaking in distilled water [18, 19].

Amide I band

The Amide I band is directly related to the secondary structure of proteins. It can be deconvoluted into several sub-bands associated with different forms of secondary structure; i.e. α helix, ß sheet, turns and random coil structures [73-75].

Generally, secondary structures of proteins are formed by the intramolecular hydrogen connections between, on one side, the oxygen of the carbonyl group and on the other side the NH group. These organized structures can be affected by the addition/remove of water molecules. The hydrogen bond (–C=O···H–N) could be weakened by an excessive presence of water molecules fitting themselves between the peptide chains (–C=O ··· H_2O ···H–N).

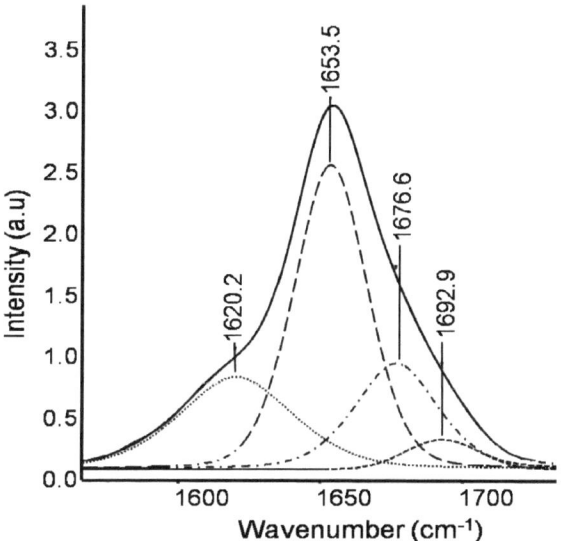

Figure 8: Deconvolution of amide I band of SC Raman spectrum. Mean spectrum at 4% RH with n=6. The fits have been done using four contributions and all parameters were free to move: 1), a band centered on 1621±5 cm-1; 2) a band centered on 1653±5 cm-1; 3) a band centered on 1678±5 cm-1; and 4) a band centered on 1690±5 cm-1.

In order to follow these variations, band deconvolution was performed on the Amide I band (figure 8). Four spectral features were identified using the second derivatives of the spectra. The first feature arising around 1621 cm^{-1} was associated in the literature with aromatic peaks of phenylalanine, tyrosine and tryptophan [45]. The second sub-band, at 1656 cm^{-1} is associated with α helix form of the keratin, while the third arising around 1678 cm^{-1} was assigned to β sheets. We have to notice that the Amide I band of SC ceramides arises at almost the same wavenumbers. Its contribution is meanwhile minimal ($\approx$1.2%). The evolution of these sub-bands can therefore be considered as mainly due to the protein secondary structure. Finally, the 1690 cm^{-1} feature was attributed to turns and random coils. The sub-band around 1621 cm^{-1} showed almost no variation while the second and the third sub-bands (1656 cm^{-1} and 1678 cm^{-1}) decreased slightly (data not shown). Finally, the intensity of the 1690 cm^{-1} sub-band (turns and random coils) increased up to 60% of RH (Figure 9) following a similar evolution compared to partially bound water (Figure 3).

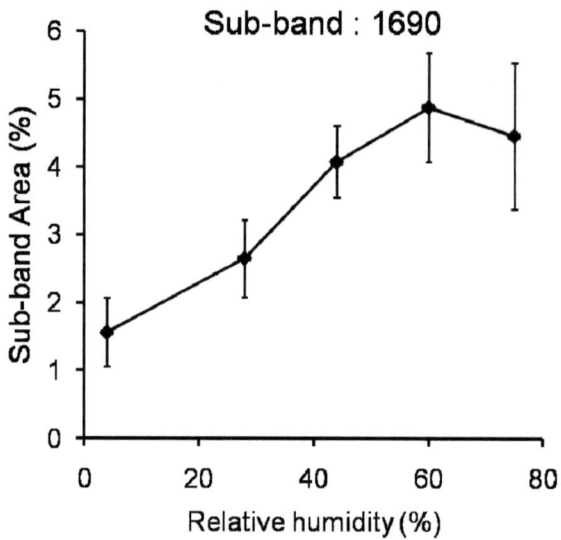

Figure 9: Evolution of an amide I fitted sub-band (centered on 1690±5 cm-1) upon humidity. Mean data at each RH level with n=6.

This is in accordance with previous studies indicating that when hydration increases, there is a continuous swelling of keratin filaments [22, 23] and unspecified changes in proteins' conformations [20]. According to our findings, these changes could correspond to the unfolding of long α helixes and β-sheets resulting in turns and random coil protein secondary structure.

Conclusion

This work shows the potential of Raman spectroscopy to evaluate the molecular and structural behavior of SC components along RH variations. It clearly demonstrates that a global understanding of the hydration/dehydration process cannot be obtained by studying only the 3350-3550 cm^{-1} spectral region. Even if this spectral feature is directly informative on the unbound water content in SC, this work proves that the study of bound water and especially the partially bound water is of great interest in the understanding of the effect of hydration on the barrier function.

Indeed, the results demonstrate that while the structure of proteins is influenced by the water content, the conformational order and the compactness of the lipid barrier in the SC are directly correlated to the variation in the partially bound water.

Furthermore, this work demonstrates that intermediate RH values provide a more organized lipid barrier. Combined with the work of Alonso et al [13] on elastic properties of the SC, suggesting a relation between the partially bound water and the elasticity parameters, our results provide a further step in the understanding of the hydration/dehydration process and the possible relation between the strain in the SC and the state of the lipid barrier function.

Such information could have a great impact in the development of new products for the prevention and correction of dermatological disorders related with low water content.

References

[1] R.R. Warner, N.A. Lilly, Correlation of water content with ultrastructure in the stratum corneum., In: Elsner P, Berardesca E, Maibach HI (eds). Bioengineering of the Skin: Water and the Stratum Corneum (1994) Boca Raton, 3-12.

[2] C. Merle, C. Laugel, A. Baillet-Guffroy, Spectral monitoring of photoirradiated skin lipids: MS and IR approaches, Chem Phys Lipids, 154 (2008) 56-63.

[3] L.D. Rhein, F.A. Simion, C. Froebe, J. Mattai, R.H. Cagan, Development of a stratum corneum lipid model to study the cutaneous moisture barrier properties, Colloids and Surfaces, 48 (1990) 1-11.

[4] J.W. Fluhr, R. Darlenski, C. Surber, Glycerol and the skin: holistic approach to its origin and functions, Br J Dermatol, 159 (2008) 23-34.

[5] M. Egawa, T. Hirao, M. Takahashi, In vivo estimation of stratum corneum thickness from water concentration profiles obtained with Raman spectroscopy, Acta Derm Venereol, 87 (2007) 4-8.

[6] M. Egawa, M. Oguri, T. Kuwahara, M. Takahashi, Effect of exposure of human skin to a dry environment, Skin Res Technol, 8 (2002) 212-218.

[7] M. Egawa, T. Kajikawa, Changes in the depth profile of water in the stratum corneum treated with water, Skin Res Technol, 15 (2009) 242-249.

[8] M. Egawa, H. Tagami, Comparison of the depth profiles of water and water-binding substances in the stratum corneum determined in vivo by Raman spectroscopy between the cheek and volar forearm skin: effects of age, seasonal changes and artificial forced hydration, Br J Dermatol, 158 (2008) 251-260.

[9] M. Boncheva, J. de Sterke, P.J. Caspers, G.J. Puppels, Depth profiling of Stratum corneum hydration in vivo: a comparison between conductance and confocal Raman spectroscopic measurements, Exp Dermatol, 18 (2009) 870-876.

[10] G.C. Charalambopoulou, T.A. Steriotis, T. Hauss, A.K. Stubos, N.K. Kanellopoulos, Structural alterations of fully hydrated human stratum corneum, Physica B: Condensed Matter, 350 (2004) E603-E606.

[11] M. Takenouchi, H. Suzuki, H. Tagami, Hydration characteristics of pathologic stratum corneum--evaluation of bound water, J Invest Dermatol, 87 (1986) 574-576.

[12] A. Alonso, N.C. Meirelles, M. Tabak, Effect of hydration upon the fluidity of intercellular membranes of stratum corneum: an EPR study, Biochim Biophys Acta, 1237 (1995) 6-15.

[13] A. Alonso, N.C. Meirelles, V.E. Yushmanov, M. Tabak, Water increases the fluidity of intercellular membranes of stratum corneum: correlation with water permeability, elastic, and electrical resistance properties, J Invest Dermatol, 106 (1996) 1058-1063.

[14] G. Imokawa, A. Abe, K. Jin, Y. Higaki, M. Kawashima, A. Hidano, Decreased level of ceramides in stratum corneum of atopic dermatitis: an etiologic factor in atopic dry skin?, J Invest Dermatol, 96 (1991) 523-526.

[15] S. Yadav, N.G. Pinto, G.B. Kasting, Thermodynamics of water interaction with human stratum corneum I: measurement by isothermal calorimetry, J Pharm Sci, 96 (2007) 1585-1597.

[16] A.V. Rawlings, Trends in stratum corneum research and the management of dry skin conditions, Int J Cosmet Sci, 25 (2003) 63-95.

[17] A.V. Rawlings, P.J. Matts, Stratum corneum moisturization at the molecular level: an update in relation to the dry skin cycle, J Invest Dermatol, 124 (2005) 1099-1110.

[18] J.A. Bouwstra, H.W. Groenink, J.A. Kempenaar, S.G. Romeijn, M. Ponec, Water distribution and natural moisturizer factor content in human skin equivalents are regulated by environmental relative humidity, J Invest Dermatol, 128 (2008) 378-388.

[19] T. Richter, J.H. Müller, U.D. Schwarz, R. Wepf, R. Wiesendanger, Investigation of the swelling of human skin cells in liquid media by tapping mode scanning force microscopy, Applied Physics A: Materials Science & Processing, 72 (2001) S125-S128.

[20] R.L. Anderson, J.M. Cassidy, J.R. Hansen, W. Yellin, Hydration of stratum corneum, Biopolymers, 12 (1973) 2789-2802.

[21] Y. Jokura, S. Ishikawa, H. Tokuda, G. Imokawa, Molecular analysis of elastic properties of the stratum corneum by solid-state 13C-nuclear magnetic resonance spectroscopy, J Invest Dermatol, 104 (1995) 806-812.

[22] C.L. Silva, D. Topgaard, V. Kocherbitov, J.J. Sousa, A.A. Pais, E. Sparr, Stratum corneum hydration: phase transformations and mobility in stratum corneum, extracted lipids and isolated corneocytes, Biochim Biophys Acta, 1768 (2007) 2647-2659.

[23] J.A. Bouwstra, A. de Graaff, G.S. Gooris, J. Nijsse, J.W. Wiechers, A.C. van Aelst, Water distribution and related morphology in human stratum corneum at different hydration levels, J Invest Dermatol, 120 (2003) 750-758.

[24] R.R. Warner, K.J. Stone, Y.L. Boissy, Hydration disrupts human stratum corneum ultrastructure, J Invest Dermatol, 120 (2003) 275-284.

[25] J.A. Bouwstra, G.S. Gooris, W. Bras, D.T. Downing, Lipid organization in pig stratum corneum, J Lipid Res, 36 (1995) 685-695.

[26] J.A. Bouwstra, P.L. Honeywell-Nguyen, G.S. Gooris, M. Ponec, Structure of the skin barrier and its modulation by vesicular formulations, Prog Lipid Res, 42 (2003) 1-36.

[27] D.A. Van Hal, E. Jeremiasse, H.E. Junginger, F. Spies, J.A. Bouwstra, Structure of fully hydrated human stratum corneum: a freeze-fracture electron microscopy study, J Invest Dermatol, 106 (1996) 89-95.

[28] G.M. Golden, D.B. Guzek, A.H. Kennedy, J.E. McKie, R.O. Potts, Stratum corneum lipid phase transitions and water barrier properties, Biochemistry, 26 (1987) 2382-2388.

[29] M. Boncheva, F. Damien, V. Normand, Molecular organization of the lipid matrix in intact Stratum corneum using ATR-FTIR spectroscopy, Biochim Biophys Acta, 1778 (2008) 1344-1355.

[30] S. Wartewig, R. Neubert, W. Rettig, K. Hesse, Structure of stratum corneum lipids characterized by FT-Raman spectroscopy and DSC. IV. Mixtures of ceramides and oleic acid, Chemistry and Physics of Lipids, 91 (1998) 145-152.

[31] M. Gniadecka, O. Faurskov Nielsen, D.H. Christensen, H.C. Wulf, Structure of water, proteins, and lipids in intact human skin, hair, and nail, J Invest Dermatol, 110 (1998) 393-398.

[32] P.J. Caspers, G.W. Lucassen, R. Wolthuis, H.A. Bruining, G.J. Puppels, In vitro and in vivo Raman spectroscopy of human skin, Biospectroscopy, 4 (1998) S31-39.

[33] P.J. Caspers, G.W. Lucassen, E.A. Carter, H.A. Bruining, G.J. Puppels, In vivo confocal Raman microspectroscopy of the skin: noninvasive determination of molecular concentration profiles, J Invest Dermatol, 116 (2001) 434-442.

[34] P.J. Caspers, G.W. Lucassen, G.J. Puppels, Combined in vivo confocal Raman spectroscopy and confocal microscopy of human skin, Biophys J, 85 (2003) 572-580.

[35] L. Chrit, P. Bastien, B. Biatry, J.T. Simonnet, A. Potter, A.M. Minondo, F. Flament, R. Bazin, G.D. Sockalingum, F. Leroy, M. Manfait, C. Hadjur, In vitro and in vivo confocal Raman study of human skin hydration: assessment of a new moisturizing agent, pMPC, Biopolymers, 85 (2007) 359-369.

[36] M. Gniadecka, O.F. Nielsen, S. Wessel, M. Heidenheim, D.H. Christensen, H.C. Wulf, Water and protein structure in photoaged and chronically aged skin, J Invest Dermatol, 111 (1998) 1129-1133.

[37] J.Q. Wu, L. Kilpatrick-Liverman, Characterizing the composition of underarm and forearm skin using confocal raman spectrscopy, International Journal of Cosmetic Science, 33 257-262.

[38] A. Tfayli, E. Guillard, M. Manfait, A. Baillet-Guffroy, Thermal dependence of Raman descriptors of ceramides. Part I: effect of double bonds in hydrocarbon chains, Anal Bioanal Chem, 397 (2010) 1281-1296.

[39] S. Raudenkolb, S. Wartewig, G. Brezesinski, S.S. Funari, R.H. Neubert, Hydration properties of N-(alpha-hydroxyacyl)-sphingosine: X-ray powder diffraction and FT-Raman spectroscopic studies, Chem Phys Lipids, 136 (2005) 13-22.

[40] M. Wegener, R. Neubert, W. Rettig, S. Wartewig, Structure of stratum corneum lipids characterized by FT-Raman spectroscopy and DSC. III. Mixtures of ceramides and cholesterol, Chem Phys Lipids, 88 (1997) 73-82.

[41] A. Percot, M. Lafleur, Direct Observation of Domains in Model Stratum Corneum Lipid Mixtures by Raman Microspectroscopy, Biophysical journal, 81 (2001) 2144-2153.

[42] E. Guillard, A. Tfayli, M. Manfait, A. Baillet-Guffroy, Thermal dependence of Raman descriptors of ceramides. Part II: effect of chains lengths and head group structures, Anal Bioanal Chem, 399 1201-1213.

[43] A. Tfayli, E. Guillard, M. Manfait, A. Baillet-Guffroy, Raman spectroscopy: feasibility of in vivo survey of stratum corneum lipids, effect of natural aging, Eur J Dermatol.

[44] G. Zeng, J.J. Shou, K.K. Li, Y.H. Zhang, In-situ confocal Raman observation of structural changes of insulin crystals in sequential dehydration process, Biochim Biophys Acta, 1814 (2011) 1631-1640.

[45] T. Lefèvre, M.-E. Rousseau, M. Pézolet, Protein Secondary Structure and Orientation in Silk as Revealed by Raman Spectromicroscopy, Biophysical journal, 92 (2007) 2885-2895.

[46] Q. Zhang, K.L. Andrew Chan, G. Zhang, T. Gillece, L. Senak, D.J. Moore, R. Mendelsohn, C.R. Flach, Raman microspectroscopic and dynamic vapor sorption characterization of hydration in collagen and dermal tissue, Biopolymers, 95 607-615.

[47] G.E. Walrafen, Y.C. Chu, Linearity between Structural Correlation Length and Correlated-Proton Raman Intensity from Amorphous Ice and Supercooled Water up to Dense Supercritical Steam, The Journal of Physical Chemistry, 99 (1995) 11225-11229.

[48] Q. Sun, The Raman OH stretching bands of liquid water, Vibrational Spectroscopy, 51 (2009) 213-217.

[49] P. Leary, F. Adar, R. Carlton, J. Reffner, F. Kang, R. Mueller, Hydration Studies of Pharmaceuticals Using IR and Raman Spectroscopy, Microscopy and Microanalysis, 13 (2007) 1694-1695.

[50] E.E. Lawson, A.N. Anigbogu, A.C. Williams, B.W. Barry, H.G. Edwards, Thermally induced molecular disorder in human stratum corneum lipids compared with a model phospholipid system; FT-Raman spectroscopy, Spectrochim Acta A Mol Biomol Spectrosc, 54A (1998) 543-558.

[51] J. Caussin, G.S. Gooris, M. Janssens, J.A. Bouwstra, Lipid organization in human and porcine stratum corneum differs widely, while lipid mixtures with porcine ceramides model human stratum corneum lipid organization very closely, Biochim Biophys Acta, 1778 (2008) 1472-1482.

[52] A. Tfayli, O. Piot, F. Draux, F. Pitre, M. Manfait, Molecular characterization of reconstructed skin model by Raman microspectroscopy: comparison with excised human skin, Biopolymers, 87 (2007) 261-274.

[53] S. Gnanakaran, R.M. Hochstrasser, A.E. GarcÃ-a, Nature of structural inhomogeneities on folding a helix and their influence on spectral measurements, Proceedings of the National Academy of Sciences of the United States of America, 101 (2004) 9229-9234.

[54] D.M. Byler, H. Susi, Examination of the secondary structure of proteins by deconvolved FTIR spectra, Biopolymers, 25 (1986) 469-487.

[55] H. Torii, M. Tasumi, Model calculations on the amide-I infrared bands of globular proteins, The Journal of Chemical Physics, 96 (1992) 3379-3387.

[56] M. Nicollier, P. Agache, J.L. Kienzler, R. Laurent, R. Gibey, N. Cardot, J.C. Henry, Action of trypsin on human plantar stratum corneum. An ultrastructural study, Arch Dermatol Res, 268 (1980) 53-64.

[57] L.C. Morejohn, J.N. Pratley, Differential effects of trypsin on the epidermis of Rana catesbeiana. Observations on differentiating junctions and cytoskeletons, Cell Tissue Res, 198 (1979) 349-362.

[58] V.S. Thakoersing, M.O. Danso, A. Mulder, G. Gooris, A. El Ghalbzouri, J.A. Bouwstra, Nature versus nurture: does human skin maintain its stratum corneum lipid properties in vitro?, Experimental Dermatology, 21 (2012) 865-870.

[59] M. Förster, M.A. Bolzinger, M.R. Rovere, O. Damour, G. Montagnac, S. BrianÃ§on, Confocal Raman microspectroscopy for evaluating the stratum corneum removal by 3 standard methods, Skin Pharmacol Physiol, 24 (2011) 103-112.

[60] K. Biniek, K. Levi, R.H. Dauskardt, Solar UV radiation reduces the barrier function of human skin, Proceedings of the National Academy of Sciences, 109 (2012) 17111-17116.

[61] S. Tfaili, C. Gobinet, G. Josse, J.-F. Angiboust, M. Manfait, O. Piot, Confocal Raman microspectroscopy for skin characterization: a comparative study between human skin and pig skin, Analyst, 137 (2012) 3673-3682.

[62] J.L. Bruneel, J.C. Lassègues, C. Sourisseau, In-depth analyses by confocal Raman microspectrometry: experimental features and modeling of the refraction effects, Journal of Raman Spectroscopy, 33 (2002) 815-828.

[63] A. Tfayli, O. Piot, M. Manfait, Confocal Raman microspectroscopy on excised human skin: uncertainties in depth profiling and mathematical correction applied to dermatological drug permeation, J Biophotonics, 1 (2008) 140-153.

[64] A. Savitzky, M.J.E. Golay, Smoothing and differentiation of data by simplified least squares procedures, Anal. Chem., 36 (1964) 1627-1639.

[65] J. Zhao, H. Lui, D.I. McLean, H. Zeng, Automated autofluorescence background subtraction algorithm for biomedical Raman spectroscopy, Appl Spectrosc, 61 (2007) 1225-1232.

[66] A. Huizinga, A.C. Bot, F.F. de Mul, G.F. Vrensen, J. Greve, Local variation in absolute water content of human and rabbit eye lenses measured by Raman microspectroscopy, Exp Eye Res, 48 (1989) 487-496.

[67] N.J. Bauer, J.P. Wicksted, F.H. Jongsma, W.F. March, F. Hendrikse, M. Motamedi, Noninvasive assessment of the hydration gradient across the cornea using confocal Raman spectroscopy, Invest Ophthalmol Vis Sci, 39 (1998) 831-835.

[68] S. Leikin, V.A. Parsegian, W.H. Yang, G.E. Walrafen, Raman spectral evidence for hydration forces between collagen triple helices, Proceedings of the National Academy of Sciences, 94 (1997) 11312-11317.

[69] J. Pieper, G. Charalambopoulou, T. Steriotis, S. Vasenkov, A. Desmedt, R.E. Lechner, Water diffusion in fully hydrated porcine stratum corneum, Chemical Physics, 292 (2003) 465-476.

[70] J.J. Bulgin, L.J. Vinson, The use of differential thermal analysis to study the bound water in stratum corneum membranes, Biochimica et Biophysica Acta (BBA) - General Subjects, 136 (1967) 551-560.

[71] S. Verdier-Sevrain, F. Bonte, Skin hydration: a review on its molecular mechanisms, J Cosmet Dermatol, 6 (2007) 75-82.

[72] W. Sun, Q. Zhao, M. Zhao, B. Yang, C. Cui, J. Ren, Structural Evaluation of Myofibrillar Proteins during Processing of Cantonese Sausage by Raman Spectroscopy, Journal of Agricultural and Food Chemistry, 59 (2011) 11070-11077.

[73] W. Akhtar, H.G.M. Edwards, Fourier-transform Raman spectroscopy of mammalian and avian keratotic biopolymers, Spectrochimica Acta Part A: Molecular and Biomolecular Spectroscopy, 53 (1997) 81-90.

[74] Anigbogu, C. A. N, Williams, C. A, Barry, W. B, Edwards, M. H. G, Fourier transform Raman spectroscopy of interactions between the penetration enhancer dimethyl sulfoxide and human stratum corneum, Elsevier, Amsterdam, PAYS-BAS, 1995.

[75] L. Chrit, C. Hadjur, S. Morel, G. Sockalingum, G. Lebourdon, F. Leroy, M. Manfait, In vivo chemical investigation of human skin using a confocal Raman fiber optic microprobe, J Biomed Opt, 10 (2005) 44007.

CHAP V. EFFETS DE L'HYDRATATION SUR LES PROPRIETES MECANIQUES DU *STRATUM CORNEUM*

ARTICLE 2: RAPPORT ENTRE LA PERTE D'EAU, LES PROPRIETES MECANIQUES ET LA STRUCTURE SUPRA-MOLECULAIRE DU SC HUMAIN ISOLE

Contexte : Une peau sèche se caractérise par une augmentation de la rugosité, de la tension et de la rigidité. Ces caractéristiques mécaniques de la peau dépendent largement de l'hydratation du SC. Néanmoins, les aspects moléculaires de la sécheresse cutanée et sa relation avec l'état mécanique du SC ne sont pas totalement élucidés.

Méthodes : Dans ce travail, le stress mécanique a été mesuré en fonction du temps de séchage sur un SC initialement hydraté (98%) puis exposé à un environnement sec (7%). Ce stress a été corrélé à des modifications structurales du SC. Les différents états de l'eau ont été quantifiés tout au long du processus de désorption d'eau. Pour une meilleure compréhension du processus d'assèchement et d'apparition du stress mécanique du SC, les modifications des descripteurs lipidiques et protéiques en fonction du temps de séchage ont été également étudiées.

Résultats : Les mesures comparatives (Raman et gravimétrie) ont permis d'établir la relation suivante : Y = 69,03 (rapport d'air sous courbes des bandes vOH/vCH) - 8.016. Cette équation permet d'évaluer directement le niveau d'hydratation du SC par spectroscopie Raman. Lors du dessèchement du SC, la compacité de la matrice lipidique du SC suit la même évolution que la quantité relative d'eau partiellement liée, et simultanément, la modification de la structure des protéines a révélé une transformation des feuillets β en hélices α. La contrainte biomécanique du SC augmente avec la perte d'eau et dépend notamment de la fraction d'eau non liée.

Conclusion : Nos résultats ont montré que le stress mécanique est lié aussi bien au niveau d'hydratation qu'à des modifications supramoléculaires des lipides et des protéines. Cette association d'information moléculaire et organisationnelle pourrait conduire, dans le futur, à la construction des modèles mathématiques permettant de déduire des informations indirectes sur le stress mécanique et l'ultra-structure des composants du SC à partir du degré d'hydratation.

ARTICLE 2

The relationship between water loss, mechanical stress, and molecular structure of human *stratum corneum ex vivo*

Raoul Vyumvuhore[1], Ali Tfayli[1], Krysta Biniek[2], Hélène Duplan[3], Alexandre Delalleau[3], Michel Manfait[4], Reinhold Dauskardt[2], Arlette Baillet-Guffroy[1]

[1]Laboratory of analytical chemistry, Analytical Chemistry Group of Paris-Sud (GCAPS-EA4041), Faculty of pharmacy, University of Paris-Sud , Chatenay-Malabry, France

[2]Department of Materials Science and Engineering, Stanford University, Stanford, CA 94305, USA

[3]Center of Research Pierre Fabre Dermo-Cosmetics (PFDC), Toulouse, France

[4]MéDyC UMR CNRS 6237, MEDIAN unit, University of Reims Champagne-Ardennes, Reims, France

Abstract

Proper hydration of the *stratum corneum* (SC) is important for maintaining skin's vital functions. Water loss causes development of drying stresses, which can be perceived as 'tightness', and plays an important role in dry skin damage processes. However, molecular structure modifications arising from water loss and the subsequent development of stress has not been established. We investigated the drying stress mechanism by studying, *ex vivo*, the behaviors of the SC components during water desorption from initially fully hydrated samples using Raman spectroscopy. Simultaneously, we measure the SC mechanical stress with a substrate curvature instrument.

Very good correlations of water loss to the mechanical stress of the stratum corneum were obtained, and the latter was found to depend mainly on the unbound water fraction. In addition to that, the water loss is accompanied with an increase of lipids matrix compactness characterized by lower chain freedom, while protein structure showed an increase in amount of α-helices, a decline in β-sheets, and an increase in folding in the tertiary structure of keratin. The drying process of SC involves a complex interplay of water binding, molecular modifications, and mechanical stress. This article provides a better understanding of the molecular mechanism associated to SC mechanics.

Key words: *stratum corneum*, skin hydration, mechanical stress, Raman spectroscopy

Introduction

Dry skin conditions, including *stratum corneum* (SC) drying stresses, are prevalent in the general population especially in aged skin[1]. Drying stresses have been well characterized as a function of the chemical potential of water in the environment as well as in the presence of molecules that can alter the barrier function of SC [2-5]. However, many questions remain about how drying stresses relate to the water content of the SC, and more specifically the various states of water within the SC as well as proteins and lipids conformational changes. These stresses can be significant and can result in chapping and cracking. Such damage can compromise the barrier function, resulting in detrimental skin reactions including inflammation, infection, scarring, abnormal desquamation and further aggravate skin disorders such as atopic dermatitis, ichthyosis vulgaris and chronic xerosis [6-9].

Water within the SC exists in a number of states including bound water which is tightly bound to the polar sites of proteins and does not vary with humidity, partially bound water which binds to the bound water and other SC molecular components, and unbound water which shares no hydrogen bonds with SC components. The complex interplay between these various states is important for maintaining the hydration of the SC. Raman spectroscopy has been shown useful for quantifying the amount of water in SC [10, 11]. Here, we use Raman spectroscopy to not only quantify the water content at specific depths in the SC tissue, but also differentiate between these water states and monitor their relative amounts as a function of time during water desorption.

Another outstanding question relates to the role of specific structural changes occurring in the SC's molecular components and their dependence on SC water content. SC hydration depends strongly on the major components including the corneocytes and their natural moisturizing factor (NMF) content, the corneodesmosomes, and the intercellular lipid matrix. Raman spectroscopy can provide detailed molecular structure information ranging from lipid conformation to protein secondary structure such as alpha helix or beta sheet formation [12, 13]. This has allowed us to compare, for the first time, the stress state of the SC to its water content and molecular structure, and to monitor how these features change as water is desorbed from the SC.

Materials and Methods

Tissue Preparation

6 human SC, acquired after plastic surgery, were obtained from the abdomen of female donors aged between 20 – 80 years. Patients had given their consent. Frozen tissue was stored at -80°C until processing. SC samples were prepared from full thickness skin using the enzymatic digest method [14-18]. Subcutaneous fat and connective tissue was removed from the skin. Epidermal tissue was separated from the dermis by immersion in a 35°C water bath for 10 min followed by a 1 min soak at 60°C and then gently separated from the dermis with tweezers. SC was subsequently detached from underlying epidermis by floating the tissue in a trypsin enzymatic digest solution at 35°C for 120 min. The tissue was then spread on water and the epidermis was gently brushed away until a clear sheet of SC was obtained. The SC was then rinsed and allowed to dry on filter paper then removed and stored at low humidity (~10-20% RH) and an ambient temperature of ~18-23°C.

Raman Spectroscopy and Gravimetric Measurements

Raman spectroscopy was performed on SC samples equilibrated at high humidity and then placed in a dry environment to study water desorption. Each sample was analyzed 3 times. A humidity controlled chamber, based on an airflow principle, was built around the microscope. Air flow was controlled by volume flow controllers (Dynaval Air, Air Liquide, Paris, France). A hygrometer/thermometer (with ± 2% accuracy) was used to measure relative humidity. To obtain the desired humidity level, dry and humid air was mixed in an appropriate ratio. The temperature of the air conditioned room was maintained at 20°C±1°C.

The water desorption analysis with all methods involved in this study, was carried out on SC samples settled onto glass cover slip as described previously [5]. As the SC is fixed on a glass slide, desorption occurs only from one side of the SC, similar to *in vivo* drying. Water desorption kinetics of hydrated SC samples in a 7% RH environment were studied. The SC samples were placed in a RH controlled chamber at 98% RH for 4 hours to allow equilibration. The humidity was then set to 7% RH followed by Raman acquisitions on the sample surface with intervals of 3 minutes for 8 hours. The same protocol was used for gravimetric measurements.

A confocal Raman microspectrometer LabRam HR Evolution (Horiba Scientific, Lille, France) was used for spectral acquisition. A video image of the sample was used for accurate positioning of the laser spot on the sample. A 633 nm He:Ne laser (Toptica Photonics, Munich, Germany) giving a 10 mW power, on the sample, was used. The 633 nm excitation wavelength was chosen because it gives a high Raman Stokes signal in both the fingerprint and in high wavenumbers region. A short focal 100X objective/ NA= 0.9 (Olympus, Tokyo, Japan) was used to focus the laser light on the surface of the sample and to collect the back scattered light. The confocal pinhole was set to 1000 μm. The in-depth spot size was around 5 μm, allowing to collect signal from a third of the sample thickness. The in-depth resolution was measured using the method described by Bruneel et al. [19, 20]. The collected light was filtered through an edge filter and dispersed with a 4 cm⁻¹ spectral resolution using a 100 μm slit and a holographic grating of 600 grooves/mm. The Raman Stokes signal was recorded with a Synapse Charge-Coupled Device detector: CCD camera (Andor Technology, Belfast, UK) containing 1024 x 256 pixels. Spectral acquisition was performed using Labspec 6 software (Horiba Scientific, Lille, France). Raman measurements were performed on the SC surface in the 400–3800 cm⁻¹ spectral range. For each scan, a 20 sec exposure time was used. All spectra were smoothed using Savitzsky-Golay algorithm on 11 points [21] and baseline corrected using an automatic polynomial function [22].

Curve Fitting

Curve fitting of the OH stretching band and Amide I (AI) band of the SC Raman spectra have been carried out using the Least Squares Fitting algorithm in Matlab (The MathWorks, Inc., Natick, Massachusetts, USA). The algorithm allows the user to identify a number of sub-bands within a spectral region using the second derivative. It then automatically adjusts the combination of bands to best fit the spectral profile. The maximum shift of the sub-band position was set to ±2 cm⁻¹, the bandwidth was fixed to 50 cm⁻¹ and 19 cm⁻¹ for OH stretching band and Amide I band respectively, while intensity was left free to adapt to the fit. The quality of the fit was estimated by the standard error and the χ^2 values.

Normalization processes were performed to correct variations occurring in global spectral intensities. For νOH sub-band quantification in the 3100-3700 cm-1 range, a prior normalization was done on the νCH vibration band and the area of each sub-band was calculated, while for the determination of the secondary structure content using the

Amide I band, the area of each component was divided by the sum of the area of all Amide I components.

Substrate Curvature

The SC stresses were measured by monitoring the curvature of a substrate onto which the SC had been adhered. The relationship between SC film stress, σ_{SC}, and elastic curvature, K, is expressed by Stoney's equation:

$$\sigma_{sc} = \frac{E_{sub}}{(1 - \nu_{sub})} \frac{h^2_{sub}}{6h_{sc}} K \qquad \text{(Equation 1)}$$

Where E_{sub}, ν_{sub} and h_{sub} are the Young's modulus, Poisson's ratio and thickness of the substrate, respectively. Initial and final SC thickness values, h_{SC}, were measured with a digital micrometer and assumed to vary linearly with time from initial stress onset till the plateau (invariant level) stress was reached. The SC was assumed to be a "thin film" compared to the substrate, an assumption that requires the product of the film biaxial modulus and thickness to be $\leq 1/80^{th}$ of the equivalent product for the substrate. This requirement is easily fulfilled here. An important advantage of this assumption is that the elastic properties of the film are not required.

To measure the curvature, fully hydrated SC was adhered as described in [5] to 22 mm x 22 mm x 177 μm borosilicate glass coverslips (Fisher Scientific, 12-541-B) with reflective Cr/Au (35 Å/465 Å) films deposited on one surface. To adhere the SC to the substrate, the SC was submersed in water for 25 min. A glass cover slip on filter paper was then submersed underneath the SC and the water was removed with pipettes, allowing the SC to settle on to the cover slip. Excess SC was removed from the edges of the cover slip using a razor blade. The SC was then placed in the substrate curvature instrument, where the temperature and humidity were controlled at 23°C and 7% RH. A scanning laser equipped with a detector measured the angle of deflection versus position on the substrate and an average curvature was calculated. After the plateau stress was reached, the sample was placed in 98%RH for 2 h to rehydrate. The drying experiment was then repeated.

In the first stage of the experiment, there is a thin layer of excess water between the SC and the substrate caused by the sample preparation process that must be depleted before the stress can begin to increase. This causes a ~2h lag time in stress onset. In the second dehydration stage, that water layer is no longer present and the stress increases immediately. The plateau stresses reached in each stage are nearly identical; indicating the hydration of the SC itself is consistent in the two stages. The second stage of the experiment is used for subsequent analysis, as this avoids the extra water that is not contained in the SC.

Results and Discussion

Investigation of SC Water Desorption

In order to quantify the water content of the SC using Raman spectroscopy, we correlated the change in mass of the SC during drying with changes in Raman peaks associated with water (Fig. 1). Placed in a dry environment (7% RH), hydrated SC samples exhibited a rapid decrease in weight. This decrease was observed predominately in the first 4 hours in the dry environment followed by a plateau (Fig.2A). From Raman spectral measurements, the νOH stretching band (3100-3620 cm^{-1}), mainly associated with total water, showed that water desorption from SC occurred predominately in the first 2 h following exposure to the dry environment and was essentially complete after 4 h (Fig. 2A), This is in accordance with previous observations of water absorption and desorption kinetics on SC [25, 26].

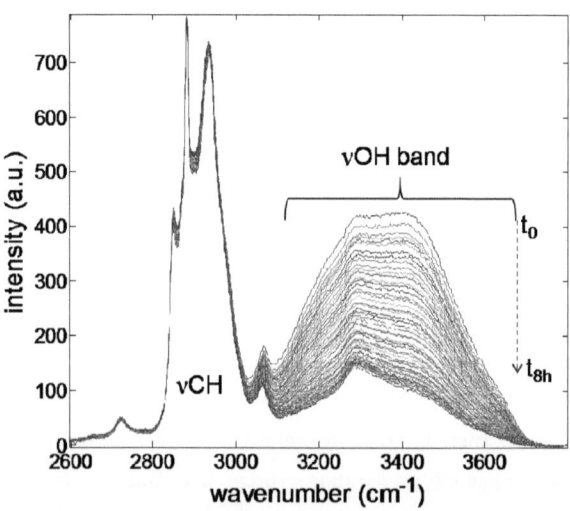

Figure 1: Raman spectra of initially hydrated SC as a function of drying time (at 7% RH): the spectral range 2600-3800 cm^{-1} shows a decrease of νOH band area.

The two methods of water quantification showed the same profile of water desorption. Meanwhile, the plateau of OH/CH ratio occurs around 4 hours in dry environment (DE), while the SC weight continues to decrease slightly down to 6.5 hours (Figure 2*A*). This could be related to the detection limit of Raman spectroscopy and/or the heterogeneity in SC water content depending on depth. Indeed, based on volume ($\sim$32 mm^3) and weight changes of the SC, the water concentration in the SC after 4hours in DE is lower than 2.5×10^{-3} mol/l which is below the detection limit of Raman [23]. Moreover, the water content may be higher in deeper layers. This could explain the difference between gravimetric measurements taking into account the whole tissue and Raman measurements collected from SC surface.

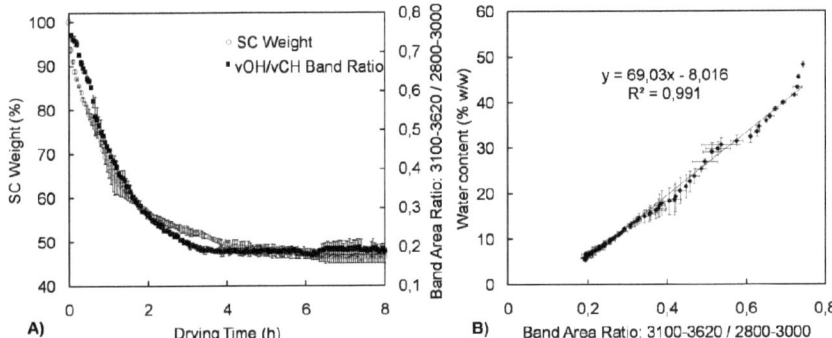

Figure 2: Quantification of water loss kinetics from SC as a function of drying time for a initially 98% RH hydrated SC exposed to 7% RH for 8 h: A) SC weight loss measured gravimetrically: percentage of weight at time "t" compared to the initial weight of the sample (at "t_o") and SC total water band area normalized on vCH band area; Water Band Ratio: 3100-3620 cm^{-1}/2800-3000 cm^{-1}, B) Correlation between gravimetric and Raman water measurements for the first 4 hours in dry environment.

We have then fitted the region corresponding to the slope of OH/CH ratio (occurring in the first 4 hours) with the equivalent water content measured by balance. A direct correlation between those two measurements was observed with a high quality of fitting; $R^2 > 0.99$ (Fig. 2*B*). We conclude that Raman spectroscopy can quantify hydration degree (in % of dry weight) by using the equation in Fig. 2*B*. However, 5% of difference between detection limit for Raman and that of balance was detected. Thus, the proposed equation in Fig. 2*B* could quantify SC water content higher than 5% of dry weight. Note that the presently used protocol allows quantification of water molecules that can be easily desorbed from SC. In contrast, the water molecules that are tightly bound to the SC will remain in the sample even at low relative humidities and are not captured by this equation.

Raman spectroscopy allows us to differentiate between different states of water, which include primary bound water, partially bound water and unbound water [27]. Primary bound water is tightly bound to the SC polar sites of proteins constituting a first monolayer of water. This type of water does not vary with relative humidity, whereas partially bound water is bound to the first monolayer and to other SC molecular components. Unbound water presents no hydrogen bonds with SC components [28, 29] . The balance between these states of water is important for skin health, appearance, and feel, and it is unknown which state of water is predominately responsible for the development of drying stresses.

Partially bound and unbound water can be observed following the AUC of the 3245-3420 cm^{-1} and 3420-3620 cm^{-1} spectral region respectively. These features were identified at equilibrium conditions by using curve fitting on the vOH spectral band (3100-3700 cm^{-1}) [27]. This allowed us to follow those types of water during the drying process. Partially bound and unbound water decrease in a dry environment following the shape of global water loss described in Fig. 2*A* (data not shown). These observations illustrate that both forms of water are lost by the SC simultaneously. This simultaneous loss of partially bound and unbound water observed here is due to the complex distribution of water in the SC [30]. At a given humidity level, some sites of SC present a single layer of water while other sites could present two or more [25, 31]. Thus, water is desorbed following slow equilibration between bound and unbound states [26]. This system is in constant modification since, as we will show, the molecular components of SC change form upon dehydration and make available buried water domains for desorption.

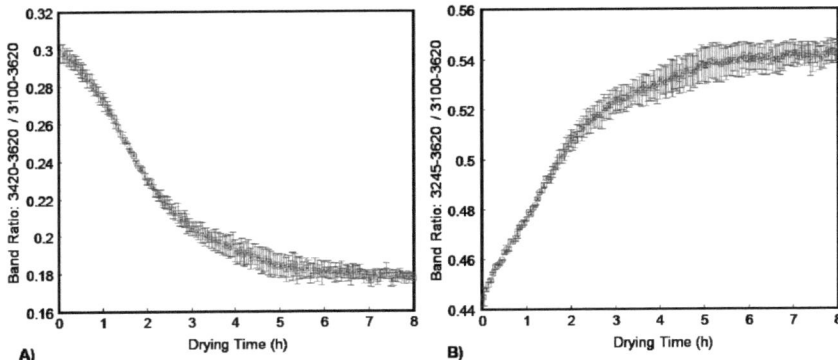

A) **B)**

Figure 3: Relative amount of partially bound and unbound water in the SC as function of drying time in dry environment (7% RH): A) Fraction of unbound water = ratio between unbound water and global water content, B) Fraction of partially bound water = ratio between partially bound water and global water content.

Despite the complexity of water distribution in the SC, we can determine how the amounts of bound and unbound water change relative to each other during the drying process. Fig. 3*A* shows that the fraction of unbound water, compared to the total water content, decreases in a dry environment while the fraction of partially bound water increases (Fig. 3*B*).

Mechanical Stress is Related to Water Content

SC stresses, σ_{SC}, were measured during water desorption from the curvature of a substrate onto which the SC had been adhered and were calculated using Eq. 1 [5, 24]. Raman spectroscopy was also performed to quantify water content of drying stress specimens as a function of time during water desorption. All analyzed samples had similar stress profiles, although plateau stresses varied due to biological variability between donors. All samples from a single donor had nearly identical drying stress profiles. For clarity of data presentation, one representative data set is illustrated in the present manuscript.

The initial stress of the hydrated SC sample was measured to be close to zero. Once the sample was placed in the dry environment, a rapid increase in stress was observed, with the stress beginning to stabilize after 2-3 h before plateau for the remainder of the time in the drying environment (Fig. 4).

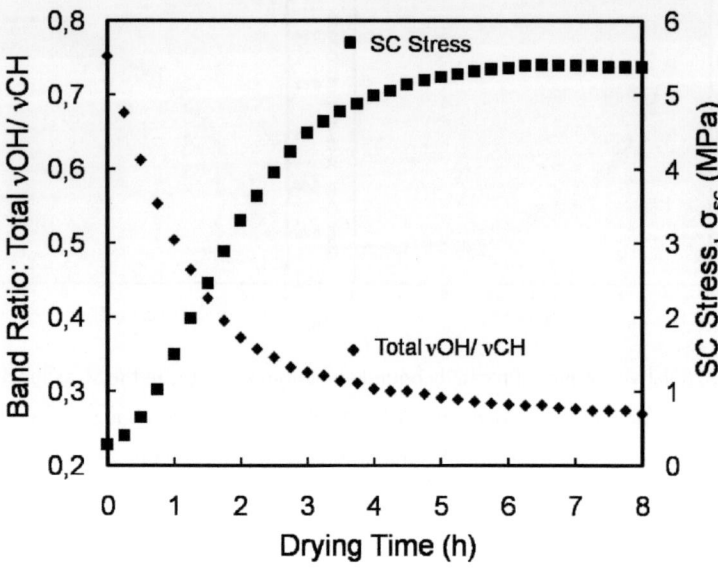

Figure 4: Comparison between total water loss and mechanical stress of the SC as a function of drying time for an initially 98% RH hydrated SC exposed to 7% RH for 8h.

The SC stress profile was compared to the evolution of water content. During the first 2h, the mechanical stress evolved inversely to the global water content (area under curve (AUC) 3100-3620 cm^{-1}) and then reached a plateau with non-significant variations. The plot of the mechanical stress against the AUC of the total vOH stretching band shows that this evolution can be divided into several steps (Fig. 4). For highly hydrated SC (first 30 min in DE), the mechanical stress is weakly affected. Then as water decreases, there is a linear increase in the stress. Finally, when the SC water content is very low, the mechanical stress is at its maximum and is constant. Although the Raman spectroscopy and drying stress measurements were performed on SC obtained from different donors, the time scale of drying is similar for all samples tested. We show representative data to illustrate general trends. Minor differences in water diffusion are expected as a result of biological variability. The correlation between drying stress and water loss is nevertheless remarkable and supports previous work showing that drying stress magnitude is related to water content [5].

An enduring question has been the degree to which the various states of water within the SC contribute to the drying stress.

We have shown that the fraction of unbound water decreases in a dry environment while the fraction of partially bound water increases (Fig. 3). Thus, we conclude that although the biomechanical drying stress of SC is due to contributions from the loss of both unbound and partially bound water, it depends mainly on the unbound water fraction. The latter is located between SC components and acts as a lubricant. Unbound water allows the SC molecules to move freely against each other and thus increases greatly the flexibility of the tissue. Since this water is removed in a dry environment, the SC molecules interact strongly, becoming more compact. This is accompanied by an increase in biomechanical stress.

Stratum Corneum Lipid Order Varies with Mechanical Stress and Water Binding

The intercellular lipid bilayers are important for controlling the movement of water through the tissue by providing the water a tortuous path of hydrophilic and hydrophobic domains through the SC. The lipid conformation is thus very important to water diffusion and hydration of skin. To study the variation of SC lipid order during the drying process, the spectral features in the $\nu C-C$ stretching region were used. The ratio between the peaks at 1060 cm^{-1} associated with *trans* conformations and the 1080 cm^{-1} feature associated with *gauche* conformations (I_{1060}/I_{1080}) is reflective of the lipid ordering (Fig. 5). High values of this ratio are associated with compact lipid packing while a decrease is indicative of looser packing, which has important implications for the SC's barrier function [32-34].

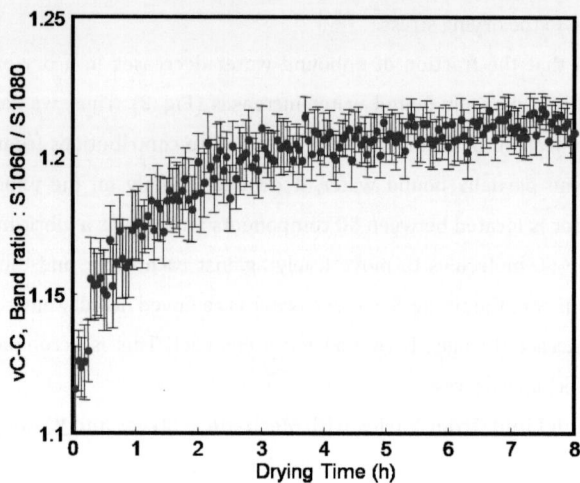

Figure 5: Evolution of lipid trans/gauche ratio (I_{1060}/I_{1080}) as function of drying time from   98% to 7% RH for 8 hours.

The *trans/gauche* ratio increased during drying (Fig. 5), reflecting lower chain freedom and higher compactness of the intercellular lipids in the SC. Lipid conformational order undergoes the same evolution as relative amount of the partially bound (Fig. 3B) water and the mechanical stress (Fig. 4). At high humidity level (t=0 min), the high number of unbound water molecules in SC reduces the intermolecular forces between the intercellular lipids. Thus, the intercellular lipids of the hydrated tissue are less tightly packed and more flexible [35]. As the tissue dehydrates, the increase in the relative amount of partially bound water (Fig. 3B) offers stronger H-bonding possibilities between water and SC molecules. In this manner, water removal could be accompanied with tight interactions between lipids involving intermolecular hydrogen bonds between lipid polar head groups and Van der Waals interactions between their hydrocarbon chains. This corroborates with previous results indicating that the fluidity of the intercellular membrane of the SC is related to the bound water content [27, 36, 37].

Protein Structure Changes with Water Desorption and Mechanical Stress

Another potential pathway for water desorption is through the corneocytes. The hydration level of the corneocytes and the conformation of the keratin matrix within is very important to the mechanical behavior of the SC as the keratin constitutes the largest volume fraction of the tissue and controls the SC's stiffness.

To study the effects of water desorption and drying stress on keratin, the Amide I band was studied. The analysis of this band gives direct information on protein secondary structure since protein contributions represent more than 98% of that band. In intact SC the α-helix conformation dominates, followed by the β-sheet, with a small amount of turns and random coil conformations. Detailed information can be obtained using curve fitting. Five spectral features were identified (Fig. 6).

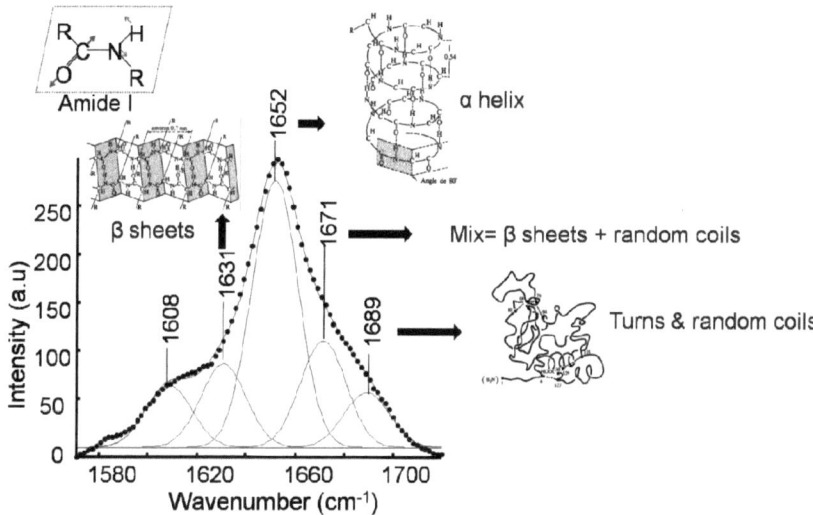

Figure 6: Deconvolution of amide I band of SC Raman spectrum. The fits have been done using five contributions; the band width was fixed to 19 cm⁻¹ and all other parameters were free to move: 1) a band centered on 1608 ± 5 cm⁻¹; 2) a band centered on 1631 ± 5 cm⁻¹; 3) a band centered on 1652 ± 5 cm⁻¹; 4) a band centered on 1671 ± 5 cm⁻¹; and 5) a band centered on 1689 ± 5 cm⁻¹.

The first feature arising around 1608 cm⁻¹ is associated with aromatic peaks of phenylalanine, tyrosine and tryptophan [38]. The second, at 1631 cm⁻¹, is attributed to β sheets, the third, at 1652 cm⁻¹ to the α helix form of the keratin, while the fourth arising around 1671 cm⁻¹ was assigned to β sheets and random coils. Finally, the 1689 cm⁻¹ feature was attributed to turns and random coils. The relative areas of the different sub-bands were calculated. The results showed mainly conversion between β-sheets and α-helixes during the drying process (Fig. 7).

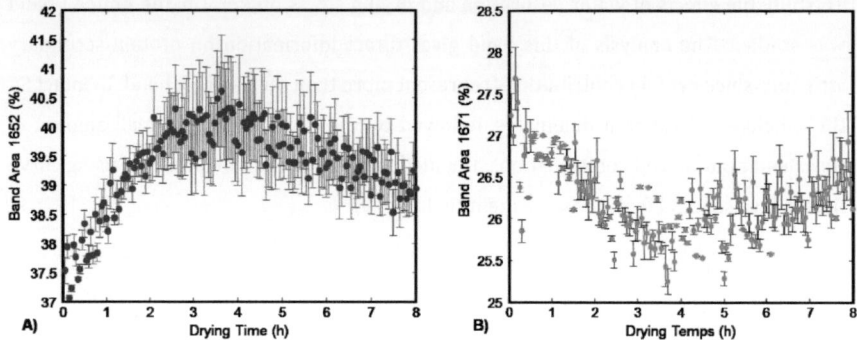

Figure 7: Evolution of protein secondary structures as function of drying time from 98% to 7% RH for 8 hours: A) α-helix structure as function of time in drying environment; B) β-sheet as function of time in drying environment.

The evolution of α-helix and β-sheet structures can be divided in several phases (Fig. 7). There was initially an increase of the amount of α-helix structures in the first 3 hours (phase I: drying process) followed by a slow decline (phase II: plateau of water loss curve), while the β-sheet structure was found to decrease in the first 3 hours and then increase slightly. The opposing behavior of these features reveals an equilibration in the secondary structure that could be related to a complex phenomenon including water loss, modification in water structure and the increase in mechanical stress.

The folding status of the keratin, i.e. tertiary structure, was also monitored by following the position of the maximum of the $v_{asym}CH_3$ band around 2932 cm⁻¹. This peak is mainly due to proteins in the SC. A Raman signal of lipid free SC is characterized by a strong $v_{asym}CH_3$ vibration around 2932 cm⁻¹ and by a decrease of vCH_2 band, while the $v_{asym}CH_3$ band in Raman signal of extracted lipid is quasi non-existent (data not shown). A shift of $v_{asym}CH_3$ towards lower wavenumbers reveals folding of the keratin thus favoring the intramolecular interactions [39, 40]. Fig. 8 shows that the folding increases with water loss and the β-sheet conformation decreases during the drying process. In phase II, the same equilibrium previously observed for the secondary structure can be observed between unfolded/folded components.

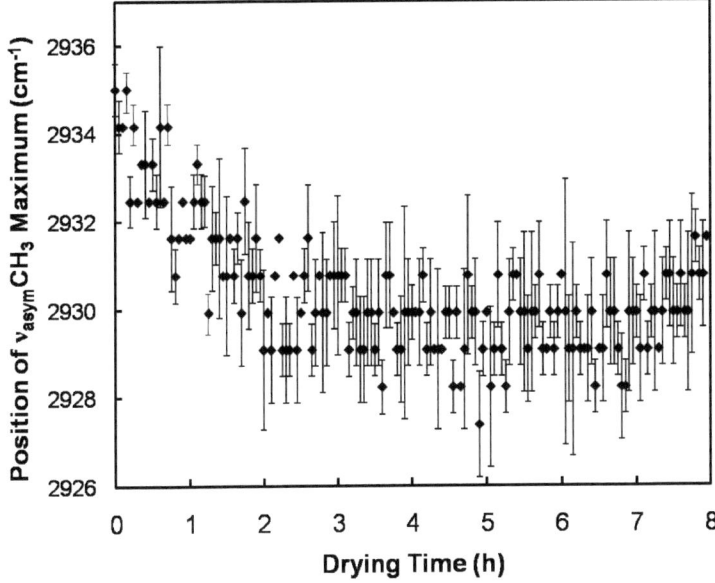

Figure 8: Protein tertiary structure as function of drying time from 98% to 7% RH for 8 hours. Maximum position of the $\nu_{asym}CH_3$ band centered around 2932 cm^{-1}.

In a highly humidified environment, the multilayer of water molecules in the SC disrupt the hydrogen bonding between the keratin chains allowing them to be less constrained and move more easily relative to each other [41, 42]. Meanwhile, the presence of water increases the volume of the SC, fills the spaces between keratin fibers and privileges β-sheet formation. The removal of this water by placing the sample in a dry environment creates spaces between keratin fibers thus reducing the possibility of β-sheet formation. Simultaneously, the loss of unbound water and the increase of the relative amount of partially bound water provide higher hydrogen bonding possibilities between C=O and N–H groups [25, 43, 44]. Thus an increase in α-helix features can be observed. Once the plateau in the water loss and mechanical stress curves is reached, a new equilibrium takes place. Under a constant mechanical stress a new rearrangement between α-helixes and β-sheets can be observed.

Conclusion

We show that water desorption from the SC has dramatic effects on the SC's drying stress, and demonstrate in particular its correlation with unbound water fraction and protein and lipid conformation. A relationship between the SC water content and its stress degree was established *ex vivo*, and we found the SC's biomechanical stress mainly depends on the unbound water fraction. We have also highlighted the protein and lipid conformational modifications due to the drying stress, showing that the compactness of the lipid matrix in the SC follows the same evolution as relative amount of partially bound water compared to the global water content, while the protein structure is modified by an increase in α-helices and a decrease in β-sheet formation. This structure modification is related to complex phenomenon including water loss, modification in water structure and an increase of the mechanical stress. Based on the analysis, we conclude that hydration degree can provide indirect information about mechanical stress and SC component structure. This is vital for understanding and treating dry skin diseases.

References

[1]   C. Paul, S. Maumus-Robert, J. Mazereeuw-Hautier, C. N. Guyen, X. Saudez and A. M. Schmitt, Dermatology, **223**, 260-265, (2011).

[2]   K. Levi and R. H. Dauskardt, Int J Cosmet Sci, **32**, 294-298, (2010).

[3]   K. Levi, A. Kwan, A. S. Rhines, M. Gorcea, D. J. Moore and R. H. Dauskardt, Br J Dermatol, **163**, 695-703, (2010).

[4]   K. Levi, A. Kwan, A. S. Rhines, M. Gorcea, D. J. Moore and R. H. Dauskardt, J Dermatol Sci, **61**, 129-131, (2011).

[5]   K. Levi, R. J. Weber, J. Q. Do and R. H. Dauskardt, Int J Cosmet Sci, **32**, 276-293, (2010).

[6]   L. E. Gaul and G. B. Underwood, J Invest Dermatol, **19**, 9-19, (1952).

[7]   C. R. Harding, S. Long, J. Richardson, J. Rogers, Z. Zhang, A. Bush and A. V. Rawlings, Int J Cosmet Sci, **25**, 157-167, (2003).

[8]   R. F. Edlich and B. A. Carl, J Emerg Med, **16**, 759-760, (1998).

[9]   E. Larson, C. Friedman, J. Cohran, J. Treston-Aurand and S. Green, Heart & Lung: The Journal of Acute and Critical Care, **26**, 404-412, (1997).

[10]   M. Egawa, T. Hirao and M. Takahashi, Acta Derm Venereol, **87**, 4-8, (2007).

[11]   P. J. Caspers, G. W. Lucassen, H. A. Bruining and G. J. Puppels, Journal of Raman Spectroscopy, **31**, 813-818, (2000).

[12]   R. Vyumvuhore, A. Tfayli, H. Duplan, A. Delalleau, M. Manfait and A. Baillet-Guffroy, Journal of Raman Spectroscopy, **44**, 1077-1083, (2013).

[13]   A. Tfayli, D. Jamal, R. Vyumvuhore, M. Manfait and A. Baillet-Guffroy, Analyst, **138**, 6582-6588, (2013).

[14]   K. Biniek, K. Levi and R. H. Dauskardt, Proceedings of the National Academy of Sciences, **109**, 17111-17116, (2012).

[15]   M. Nicollier, P. Agache, J. L. Kienzler, R. Laurent, R. Gibey, N. Cardot and J. C. Henry, Arch Dermatol Res, **268**, 53-64, (1980).

[16]   L. C. Morejohn and J. N. Pratley, Cell Tissue Res, **198**, 349-362, (1979).

[17]   V. S. Thakoersing, M. O. Danso, A. Mulder, G. Gooris, A. El Ghalbzouri and J. A. Bouwstra, Experimental Dermatology, **21**, 865-870, (2012).

[18]   M. Förster, M. A. Bolzinger, M. R. Rovere, O. Damour, G. Montagnac and S. Briançon, Skin Pharmacol Physiol, **24**, 103-112, (2011).

[19]   J. L. Bruneel, J. C. Lassègues and C. Sourisseau, Journal of Raman Spectroscopy, **33**, 815-828, (2002).

[20]   A. Tfayli, O. Piot and M. Manfait, J Biophotonics, **1**, 140-153, (2008).

[21]   A. Savitzky and M. J. E. Golay, Anal. Chem., **36**, 1627-1639, (1964).

[22]   J. Zhao, H. Lui, D. I. McLean and H. Zeng, Appl Spectrosc, **61**, 1225-1232, (2007).

[23]   T. D. Nguyen Hong, M. Jouan, N. Quy Dao, M. Bouraly and F. Mantisi, Journal of Chromatography A, **743**, 323-327, (1996).

[24]   G. G. Stoney, The tension of metallic films deposited by electrolysis, 1909.

[25]   S. Yadav, N. G. Pinto and G. B. Kasting, J Pharm Sci, **96**, 1585-1597, (2007).

[26]   Y. G. Anissimov and M. S. Roberts, Journal of Pharmaceutical Sciences, **98**, 772-781, (2009).

[27]   R. Vyumvuhore, A. Tfayli, H. Duplan, A. Delalleau, M. Manfait and A. Baillet-Guffroy, Analyst, **138**, 4103-4111, (2013).

[28]   J. A. Bouwstra, A. de Graaff, G. S. Gooris, J. Nijsse, J. W. Wiechers and A. C. van Aelst, J Invest Dermatol, **120**, 750-758, (2003).

[29]   J. J. Bulgin and L. J. Vinson, Biochimica et Biophysica Acta (BBA) - General Subjects, **136**, 551-560, (1967).

[30]   S. Verdier-Sevrain and F. Bonte, J Cosmet Dermatol, **6**, 75-82, (2007).

[31]   S. Brunauer, P. H. Emmett and E. Teller, Journal of the American Chemical Society, **60**, 309-319, (1938).

[32]   P. J. Caspers, G. W. Lucassen, R. Wolthuis, H. A. Bruining and G. J. Puppels, Biospectroscopy, **4**, S31-39, (1998).

[33]   A. Tfayli, E. Guillard, M. Manfait and A. Baillet-Guffroy, Anal Bioanal Chem, **397**, 1281-1296, (2010).

[34]   E. Guillard, A. Tfayli, M. Manfait and A. Baillet-Guffroy, Anal Bioanal Chem, **399**, 1201-1213, (2011).

[35]   S. Yadav, R. R. Wickett, N. G. Pinto, G. B. Kasting and S. W. Thiel, Skin Res Technol, **15**, 172-179, (2009).

[36]   A. Alonso, N. C. Meirelles and M. Tabak, Biochim Biophys Acta, **1237**, 6-15, (1995).

[37]   A. Alonso, N. C. Meirelles, V. E. Yushmanov and M. Tabak, J Invest Dermatol, **106**, 1058-1063, (1996).

[38]   T. Lefèvre, M.-E. Rousseau and M. Pézolet, Biophysical journal, **92**, 2885-2895, (2007).

[39]   M. Gniadecka, O. Faurskov Nielsen, D. H. Christensen and H. C. Wulf, J Invest Dermatol, **110**, 393-398, (1998).

[40]   M. Gniadecka, O. F. Nielsen, S. Wessel, M. Heidenheim, D. H. Christensen and H. C. Wulf, J Invest Dermatol, **111**, 1129-1133, (1998).

[41]   R. Marks, P. A. Payne, M. Takahashi, K. Kawasaki, M. Tanaka, S. Ohta and Y. Tsuda, in Bioengineering and the Skin, Springer Netherlands, 1982), pp. 67-73.

[42]   K. S. Wu, W. W. van Osdol and R. H. Dauskardt, Biomaterials, **27**, 785-795, (2006).

[43]   Y. Jokura, S. Ishikawa, H. Tokuda and G. Imokawa, J Invest Dermatol, **104**, 806-812, (1995).

[44]   A. Alonso, J. Vasques da Silva and M. Tabak, Biochimica et Biophysica Acta (BBA) - Proteins and Proteomics, **1646**, 32-41, (2003).

CHAP VI. DETERMINATION DE DESCRIPTEURS DU STRESS MECANIQUE SUR LE *STRATUM CORNEUM* : SPECTROSCOPIE RAMAN ET PLS

ARTICLE 3 : LA SPECTROSCOPIE RAMAN : UN OUTIL POUR LA CARACTERISATION DES PROPRIETES BIOMECANIQUES DU SC

Contexte: Les propriétés mécaniques du SC constituent l'un des aspects majeurs du fonctionnement et de l'apparence de la peau. Différents phénomènes physiologiques qui affectent la structure du SC peuvent induire un stress mécanique générant ainsi des sensations d'inconfort et de tiraillement. De nombreuses études ont été réalisées sur les propriétés de traction du SC. Il est bien établi que l'extensibilité du SC est liée à sa teneur en eau, et pourtant, peu de travaux ont tenté d'élucider le mécanisme moléculaire et les modifications supramoléculaires de la matrice lipidique et des protéines au cours de l'étirement du SC. Le principal objectif de cette étude était d'évaluer au niveau moléculaire le comportement mécanique viscoélastique du SC en utilisant une technique optique non invasive couplée à l'analyse multivariée de données spectrales.

Méthodes: Dans ce travail, une déformation uni-axiale a été appliquée sur les échantillons de SC avec un pas d'allongement de 0,5% par étape. Les échantillons sont maintenus au même niveau de traction lors de l'acquisition de spectres Raman. L'extension totale observée à la rupture a été exprimée en pourcentage de la longueur initiale.

Pour l'analyse des données, deux approches ont été suivies: la première utilisant l'analyse « Partial Least Square » (PLS) afin de prédire le niveau de tension de la peau à partir des données spectroscopiques Raman ; la seconde examinant en détails la relation entre les contraintes mécaniques et la structure et la fonction du SC. Pour mettre en évidence les mécanismes moléculaires associés à la mécanique du SC, nous avons étudié les changements de conformation des lipides intercellulaires et la structure des protéines en fonction de l'extension du SC par la spectroscopie Raman.

Résultats: Nous avons développé un modèle PLS qui permet de prédire l'état de la traction du SC à partir des spectres Raman mesurés. La forte relation linéaire ($R2 = 0,990$) entre l'allongement mesuré et prédit de l'échantillon montre une forte corrélation entre le niveau de traction du SC et sa signature Raman globale. Ce résultat suggère que les modifications conformationnelles des protéines et des lipides pourraient fournir des informations sur l'état de déformation de l'échantillon. D'une part, nous avons montré que, lorsque le SC est étiré, les protéines sous forme d'hélice α se dénouent et sont converties en formes feuillets-β. D'autre part, la conformation gauche des lipides, qui correspond à la matrice lipidique moins ordonnée, augmente avec la traction. Ainsi, ces résultats suggèrent que la traction du SC tend à affaiblir la fonction barrière de la peau.

Conclusion: Dans ce travail, nous avons pu évaluer les propriétés biomécaniques de la peau de manière objective, la méthode pourrait permettre la détection des changements subtils dans la matrice lipidique et la conformation des protéines et par conséquent dans la morphologie et fonction de la peau. Il est intéressant de voir que les caractéristiques spectrales Raman sont bien corrélées avec l'état d'extension du SC et il est possible de déterminer le degré d'extension avec un spectre Raman. D'autres expériences doivent être réalisées *in vivo* et pourraient être utilisées dans l'avenir pour les applications cliniques telles que le diagnostic du degré d'inconfort de la peau et l'évaluation du comportement élastique. Cet outil pourrait également être utilisé pour évaluer l'efficacité des produits dermo-cosmétiques qui sont développés pour améliorer le comportement mécanique de la peau, non seulement pour des raisons esthétiques, mais aussi pour maintenir les conditions de l'homéostasie cutanée. Ainsi, la spectroscopie Raman semble être un outil non invasif prometteur qui peut introduire une nouvelle compréhension des aspects moléculaires des propriétés mécaniques.

ARTICLE 3

Research article

Received: 13 March 2013   Revised: 6 May 2013   Accepted: 7 May 2013   Published online in Wiley Online Library

(wileyonlinelibrary.com) DOI 10.1002/jrs.4334

Raman spectroscopy: a tool for biomechanical characterization of Stratum Corneum

Raoul Vyumvuhore,[a] Ali Tfayli,[a]* Hélène Duplan,[b] Alexandre Delalleau,[b] Michel Manfait[c] and Arlette Baillet-Guffroy[a]

[1]Laboratory of analytical chemistry, Analytical Chemistry Group of Paris-Sud (GCAPS-EA4041), Faculty of pharmacy, University of Paris-Sud , Châtenay-Malabry, France

[2]Center of Research Pierre Fabre Dermo-Cosmetics (PFDC), Toulouse, France

[3]MéDyC UMR CNRS 6237, MEDIAN unit, University of Reims Champagne-Ardennes, Reims, France.

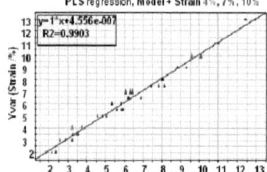

The effect of uniaxial mechanical extension on SC components ultra structure was examined. Using combination of Raman spectra and Partial Least Squares processing technique, we could determine the strain status of the analyzed samples. Furthermore, this method allowed to highlight the modifications in the protein structure and intercellular lipids bilayer organization.

Abstract:

The mechanical properties of the *stratum corneum* have been studied by different authors at the macroscopic level, but the modification of its ultra structure during mechanical extension remains unknown. Moreover, little is described about the effect of the mechanical stress on *stratum corneum* barrier function. In this study, we have examined the *stratum corneum* structure changes, at the molecular level, during uniaxial tensile experiments. This was performed on isolated *stratum corneum* samples using Raman spectroscopy. We could identify the strain status of the analyzed samples by using combination of Raman spectra and Partial Least Squares processing. In addition, this approach provided information about lipids and proteins behavior during the sample extension. The structure of the intercellular lipids bilayer became less organized up to ~9% deformation. For higher strains, a plateau corresponding to the minimum organization is observed till the complete failure of the sample. In the same time, protein structures including desmosomes, were characterized by monotonic secondary structure modifications for deformations up to ~ 9 % followed by a plateau. These observations are relevantly demonstrating the effect of extension on the skin barrier state. Such an approach could be objectively used for clinical applications to evaluate skin discomfort degree and skin elastic behavior. This could therefore help with proof of efficacy for cosmetic and dermatologic products.

Keywords: Raman spectroscopy; Partial Least Square; skin mechanical properties; skin barrier function; *stratum corneum*.

Introduction

Stratum Corneum (SC) is the most superficial layer of the epidermis. Its mechanical property stands for one of the major aspect for the skin functionality and appearance [1]. The skin mechanics depends on combination of mechanical properties of different skin layers (epidermis, dermis and hypodermis). Numerous studies on the skin have been performed to isolate the mechanical properties of the individual skin layers [2, 3]. Up to date, many researchers assume that the skin elasticity depends on the dermis and most studies were focused on this layer [4-7]. However, despite the small thickness of the SC compared to the whole skin (one hundredth); the SC modulus is much higher than that of dermis and is roughly 10 to 100 times higher than the entire skin modulus [3, 8-10]. Moreover, studying the mechanical properties of epidermis, Geerlig et al. did not find any significant difference in stiffness between the SC and viable epidermis [11]. More recently, Leveque et al. have illustrated the strong influence of SC on the global skin mechanical properties by using a physical model [1]. According to these observations, it is thought that improving the mechanical properties of the SC should improve those of the entire skin. It is now well known that the mechanical behavior of

SC is influenced by individual factors such as age [12-14], sex and health. It is also affected by environmental factors [15-17] and extraction of lipids[18]. Meanwhile, the mechanical behavior at the molecular level is not well established.

For a meaningful interpretation of the mechanical behavior of the SC, it is essential to know its composition: the SC is composed of corneocytes, which are hexagonal flat cells filled with keratin, without nucleus, held together by corneodesmosomes and lipids [19, 20]. The lipids are arranged in lamellar sheets, which consist of lipid matrix composed mainly of ceramides, cholesterol, and fatty acids. Corneodesmosomes are specialized inter-corneocyte linkages formed by proteins and, together with the lipids, they maintain the integrity of the SC [21-23]. Different physiological phenomena affecting the SC structure can induce skin stress which generates feelings of discomfort and tightness. A good understanding of the SC mechanical behavior at its molecular level is therefore essential in order to develop effective products.

For biomechanical investigations, extension [24], torsion, suction [25, 26], compression or indentation [27] techniques can be used. In the present work, only the tensile behavior was investigated. It's well established that SC extensibility is related to its water content [28]; however the relationship between the structure of SC components and its elasticity is poorly understood. It has long been thought that SC extensibility is influenced mainly by intercellular lipids and corneodesmosomes [29-33]. However, Wildnauer et al. [29] showed that extension to failure of a fully hydrated SC (100% RH) occurs at approximately 190% extension, while it decreases to 22% at 32%RH. Such differences cannot be explained only by changes in corneodesmosomes and lipids but also by modifications in corneocytes elastic properties [28]. One can assume that, the modifications in tensile properties of the SC are due to the combination of all SC component structure.

Many studies have investigated the SC tensile properties. Nevertheless, few studies have attempted to elucidate the molecular mechanism of SC lipid matrix and protein structure modifications during the extension of the SC. Some skin pathologies are characterized by combined feelings (discomfort, tightness) and a disrupted skin barrier [34-36]. Thus, the relationship between the SC strain state and its barrier function, evaluated through the conformational changes of the intercellular lipids, could give a better understanding of some skin physio-pathological status. All these characteristics could provide detailed and precise information about the stress state of the tissue.

According to the fact that a biomechanical state affects many Raman features, a multivariate regression model could better describe the SC strain levels. Partial least-squares (PLS) [37, 38] has been established as a standard data analysis tool for multivariate data analysis in the last decades and there are numerous applications in quantification and identification [39]. Therefore, PLS coupled with Raman spectroscopy could be a powerful tool for skin stress evaluation.

In the present study, two approaches were followed: the first one implementing PLS analysis in order to predict the skin strain level based on Raman spectra; the second one examining in details the relationship between mechanical strain and the SC structure and function (i.e. lipid matrix and proteins). To highlight the molecular mechanism associated to SC mechanics, we investigated molecular modifications due to the extension of SC by using Raman spectroscopy. The PLS model of the obtained spectra predicted the extension level of the tissue with a good accuracy. Thus, Raman spectroscopy appears to be a promising non-invasive tool that can introduce new comprehension of the molecular aspects of the mechanical properties.

Materials and methods

Sample preparation

Human abdominal skin samples from 5 female donors aged between 20 and 30 years were obtained after plastic surgeries and stored at -20°C. SC samples were prepared using enzymatic digestion method [40-44]. The subcutaneous fat and connective tissues were removed from the skin sample. Then, the skin tissue was immersed in distilled water (Milli-Q reagent water system) at 60°C for 1 minute. The epidermis was subsequently removed from the dermis. The epidermis was then placed, SC side up, onto a filter paper imbibed with a 0.2% Trypsin solution (0.2% in distilled water; Sigma-T4665) for 1 hour. The SC was then loosened from the filter paper using pliers. The piece of SC was spread in a water container at ambient temperature and was settled onto a greaseproof paper. The isolated SC was dried under vacuum in desiccators (containing P_2O_5).

Experimental set-up

Tensile behavior experiments were performed on 5x60mm^2 SC sheets (5x50mm^2 testing area). As shown in fig.1, the sheets were attached between the motorized stage and a stationary support. The SC samples were allowed to equilibrate 4 hours in controlled environmental conditions before measurements (50±1 % RH and 21°C±1°C).

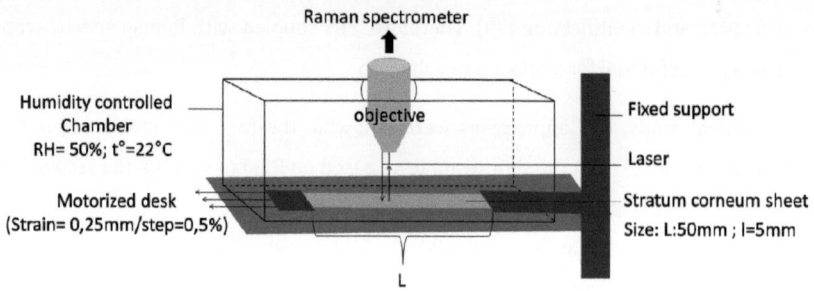

Figure 1: Experimental assembly for tensile test and Raman spectra measurement in controlled environmental chamber.

The uniaxial strain was applied on the SC samples with a step by step elongation of 250 μm/step and samples were maintained at each extension level during the Raman spectra acquisition. The total extension observed at failure was reported and was expressed in percentage of the initial length. After the tensile experiment, SC edges were analyzed to ensure that the sample did not slide during testing.

Raman microspectrometer

Samples were placed in a humidity controlled chamber under the microscope interfaced to a confocal Raman microspectrometer LabRam (Horiba Scientific, Lille, France). A video image of the sample was used for accurate positioning of the laser spot on the sample. A 660 nm pumped Nd:YLF laser (TOPTICA PHOTONICS, Munich, Germany) with a 10 mW power, on the sample, was used. The 660 nm excitation wavelength was chosen because it gives a weak fluorescence background in the fingerprint region [45] and a high Raman Stokes signal in the high wavenumbers region. A long focal microscope objective PL Fluotar L 100X/ NA 0.75 WD 4.7 (Leica, Mannheim, Germany) was used to focus the laser light on the surface of the sample and to collect the back scattered light. Confocal pinhole was set to 150 μm. The in-depth spot size was around 8 μm; the in-depth resolution was measured using the method described by bruneel et al. [46, 47]. The collected light was filtered through a notch filter and dispersed with a 4 cm⁻¹ spectral resolution using a holographic grating of 950 grooves/mm. The Raman Stokes signal was recorded with a Charge-Coupled Device detector: CCD camera (Andor technology, Belfast, UK) containing 1024 x 256 pixels. Spectral acquisition was performed using Labspec 4.18 software (Horiba Scientific, Lille, France). Raman Spectra were recorded for each step of the sample stretching. 3 measurements per step were performed for each sample. For each scan, a 20 sec exposure time was used with single accumulation. The acquisition of a spectrum lasted 90s and the recording time between 2 steps was adjusted to 5 min.For peak comparisons, Raman spectra were analyzed with software developed in house, that operates in the Matlab environment (The MathWorks, Inc., Natick, Massachusetts, USA). All spectra were smoothed using savitsky Golay algorithm on 11 points [48] and baseline corrected using an automatic polynomial function [49].

Chemometric analysis

Multivariate data analysis was performed on Simca P-11 (Umetrics, Sweden) software. Partial Least Squares (PLS) method was applied for the quantification of sample extension level and identification of discriminating wavenumbers. The X matrix of the PLS models contained smoothed and baseline corrected spectral data. The X matrix was scaled using standard normal variate (SNV) method; the Y matrix contained the relative extension levels of the tissue.

Orthogonal Signal Correction (OSC) [50] was used prior to Partial least square (PLS) regression.

Results and Discussion

Modeling strain status using SC Raman signal

PLS is a multilinear calibration method that captures the variance and correlation between X and Y; in other words, it calculates the maximum covariance between two matrix X and Y. Thus, the main aim of PLS is to predict the Y-variables from the X-variables [37-39]. Applied on Raman spectra (X-variables), the aim is to obtain an estimation of the extension level (Y-variables) from the spectral data set. The PLS decomposition summarizes the variance in the data sets in new latent variables, scores and loadings. Orthogonal Signal Correction (OSC) filtering enabled to reduce the spectral random variations. The PLS model was thus obtained with a correlation coefficient R2=0.990 and a Q2=0.915 where Q2 is the predictive ability of the model.

In order to obtain an estimate of the predictive ability of the present PLS regression model, a cross validation [51, 52] was used. The data were spit in two separate groups: a validation set containing data of 3 different levels of extension (4%, 7%, and 10%) and a calibration set containing the remaining observations. The regression model was calculated from the calibration set. The 3 extension values from validation set were then predicted using the constructed calibration model (Fig. 2). The difference between the predicted extension level and true value was very small. Root Mean Square Error of Estimation (RMSEE) and Root Mean Square Error of Prediction (RMSEP) were found to be 0.310 and 0.565, respectively. The mean confidence interval for the test set was only 0.102, demonstrating a good precision and accuracy for this model. Thus, we showed that we can quantify the extension level of a sample using its Raman spectra.

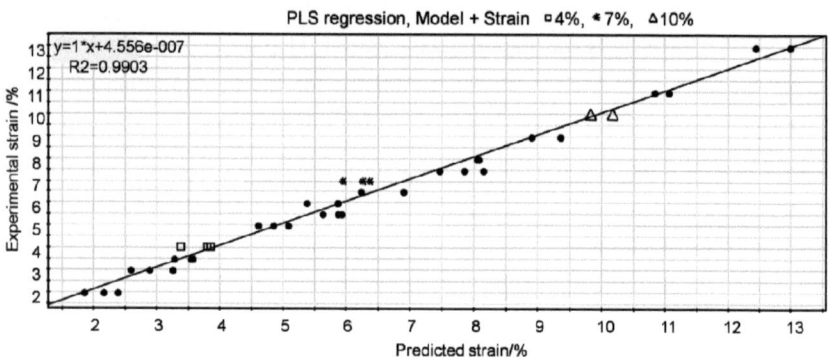

Figure 2: PLS regression after Orthogonal Signal Correction (OSC). Observed extension versus predicted extension.

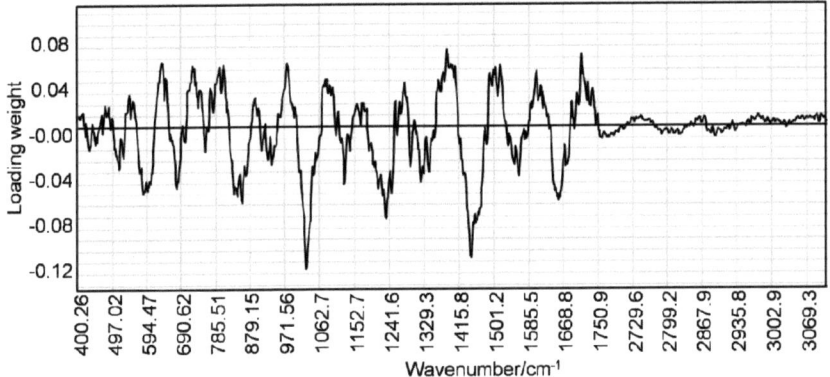

Figure 3: Loading plot of PLS regression after Orthogonal Signal Correction (OSC).

In addition to that, the loadings plot (Fig. 3) showed a large number of characteristic discriminating bands, i.e Raman features that are sensitive to the SC deformation.Those peak were similar to those assigned by visually comparing the spectra (Fig. 4). The model also enables the detection of characteristic Raman modifications which were not detected visually. This observation illustrates that informative Raman bands can be elucidated from the loadings plot of the multivariate data analysis. Hence, a proceeding like the PLS modeling detects the most important characteristic wavenumbers that are often very difficult and ambiguous to assign visually.

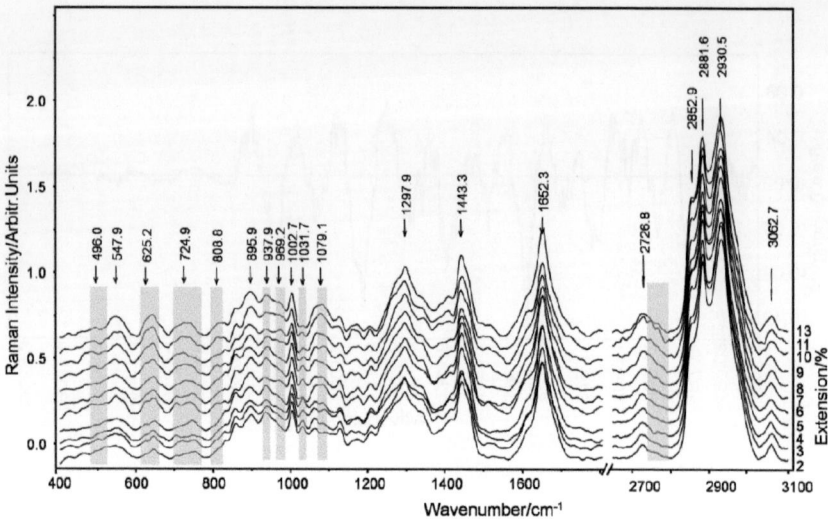

Figure 4: Mean Raman spectra of *stratum corneum* (n=3) at different strain levels. Spectral range between 400-1800 cm⁻¹ and 2650-3100 cm⁻¹ are normalized on amid I band (1565 -1720 cm⁻¹) and on C-H stretching bands (2800-3030 cm⁻¹) respectively.

Among the discriminate wavenumbers, some of them are usually used to characterize the SC protein and lipid conformational order. Thus, these features were analyzed in details to describe the SC structure modifications due to strain. Under mechanical stress, the evolution of SC components structure can be split in two separate stages. The initial slope of the strain curve is observed up to 9% of extension for lipids and for proteins, followed by a plateau till 13% deformation and sample failure (Fig. 5, 6). The same behavioral scheme was observed for all Raman features used to analyze the lipid and protein structures.

Effect of extension on SC lipid structure

Different spectral features were used in order to study the lipid conformational order in the SC: The νC–C vibration modes between 1040 cm^{-1} and 1145 cm^{-1} were used to analyze conformational changes of lipids hydrocarbon chain. The *trans/gauche* conformers ratio were analyzed by calculating $I_{1130+1060}/I_{1080}$ ratio (Fig. 5A). High values of this ratio are associated with a compact state in the lipid packing while a decrease is indicative of a loosening [53-55]. The $\nu_{asym}CH_2$ (2882 cm^{-1})/$\nu_{sym}CH_2$ (2852 cm^{-1}) ratio is generally used as an indicator of the conformational state and the lateral packing [56, 57] (Fig. 5B). High values are associated with higher *trans* content and a compact organization.

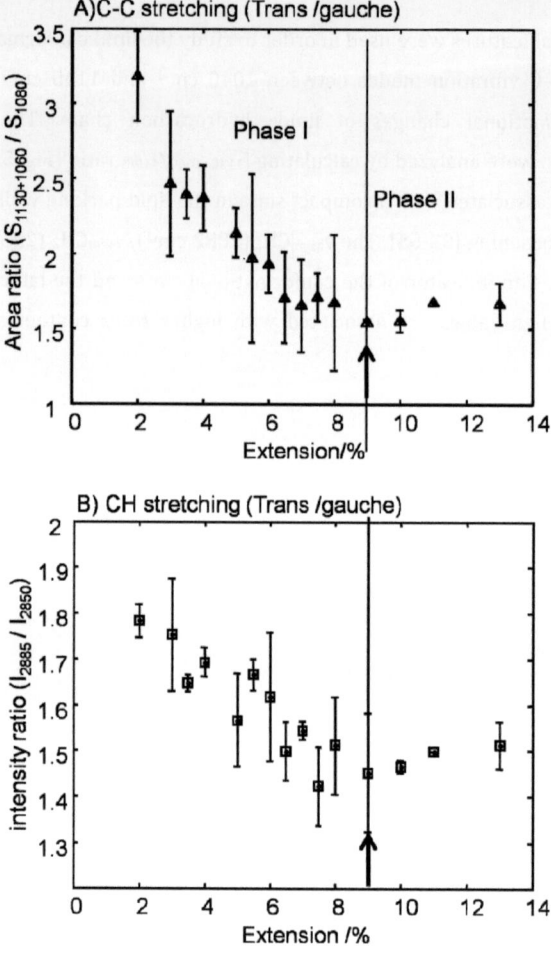

Figure 5: Conformation of *stratum corneum* lipids as function of strain status. The conformational changes were estimated using the following descriptors: A) Band area of the νC-C trans/gauche ratio= $S_{1130+1060}/ S_{1080}$; B) The νCH2 trans/gauche ratio= I_{2885} / I_{2850}. The evolution of those features could be separated in two phases.

As illustrated in Fig. 5, the *trans/gauche* ratio decreased with the extension and then stopped to vary at high extension levels (> 9 % of extension) up to failure.

Intercellular lipid structures became less organized up to 9% extension (see Fig. 5 A,B), but beyond this value, the descriptors of the lipids bilayer structure were stabilized at a low level suggesting a more "loosely packed" structure. This is in accordance with the literature assuming that during mechanical extension of the SC, a loss of lipid organization structure occurs with concomitant disturbance of water barrier function [32]. These structural changes were also detected in protein components.

Effect of extension on SC protein structure

In order to characterize the effect of SC extension on the protein structure, the evolution of the C-S vibrations and the ν C-C skeletal keratin band has been studied.

ν C–S modes of protein components arise around 700 cm^{-1} [58-61]: gauche/trans conformers ratio was evaluated by calculating I_{690}/I_{750} ratio.

The conformation of C-S-S-C group in a protein is in equilibrium of *gauche* and *trans* for each bond [62, 63]. The gauche-gauche-gauche conformation is the most stable with minimum energy. When one of the gauche becomes *trans* the stability of the structure decreases [64]. Fig. 6A showed that the gauche/trans ratio increased with strain indicating that some trans C-S are transformed in *gauche* conformation. This suggests that strain provides to the SC a more rigid structure to proteins. Indeed, the gauche conformation is mainly found in hair and nail characterized by big mechanical resistance of their proteins [65].

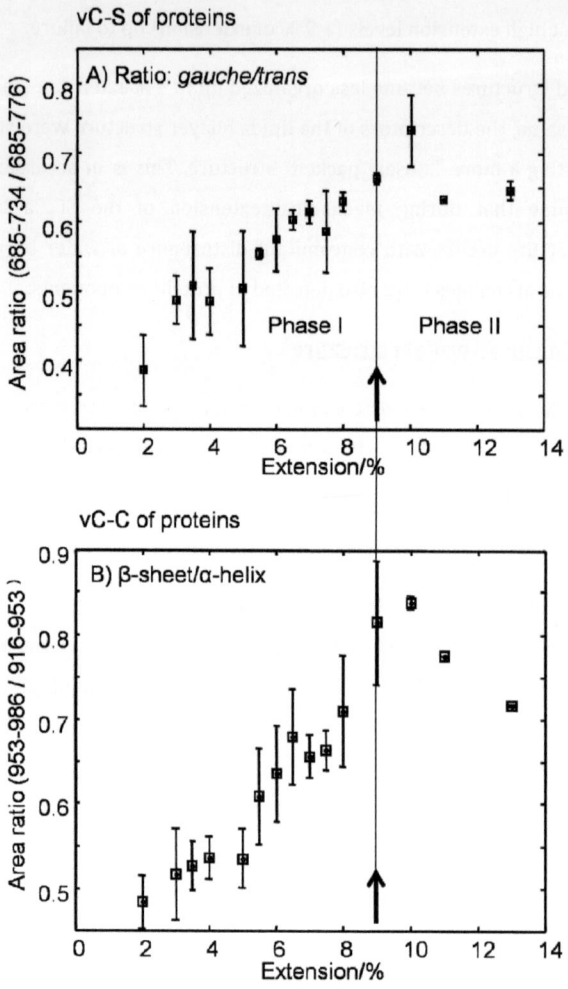

Figure 6: Conformation of *stratum corneum* protein as function of strain status. The conformational changes were estimated using the following descriptors: A) gauche/trans ratio of vC-S bands; B) β-sheet/α-helix ratio: vC-C bands of skeletal backbone of proteins.

The protein secondary structure was examined through analysis of the C-C stretching band of the skeletal keratin backbone. The band arising around 936 cm^{-1} was assigned to α-helix and the 960 cm⁻¹ feature was attributed to β sheets [60, 66, 67]. The ratio between those sub-bands was calculated (Fig. 6B). The proposed results show that the sample extension is associated with the conversion of a fraction of α-helixes into β-sheets up to 9% and then the β-sheets content slightly break down.

In the present experimental conditions, the structure of the intercellular lipid lamellae and protein are modified for strains up to 9% (Fig. 5, 6). For such environmental conditions, this could correspond to a change in the SC mechanics, from an elastic behavior to a plastic or failure stage. Indeed, for strains above 9% extension, the SC component structure remains constant at a level which differs from the initial state. Under these conditions, lipid matrix and proteins appeared disrupted comparing to their original structures. The plateau of disruption observed above 9% elongation could correspond to an elastoplastic plateau [32, 68, 69]. Nevertheless, this should be confirmed by loading and unloading experiments performed under and above 9% deformation.

According to the loading plots, except the Raman features, mentioned above describing lipids and protein conformational order, other bands sensitive to strain are located essentially around the following wavenumbers (cm⁻¹): 1030, 2700, 3060, 1650 cm⁻¹.

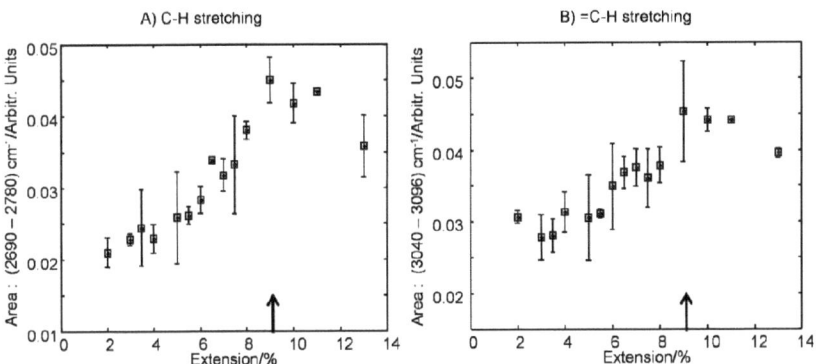

Figure 7: Intensity changes of C-H stretching bands as function of tissue extension: A) -C-H stretching bands mostly in proteins; B) Olefinic =C-H stretching vibrations.

The C-H stretching band and =C-H stretching bands arising around 2700 cm⁻¹ [70] and 3060 cm⁻¹ [71, 72] respectively are affected by strain (Fig. 7). These bands move in the same way as features described above (Fig. 5 and 6). Since those bands are not usually affected by other physiological phenomena, they could be used as specific Raman descriptors of the SC strain. The C- C cis stretching of protein around 1030 cm^{-1} [73] is one of the most important features that contribute to classify samples according to their extension level. This band starts to decreases after 5% of tissue extension and could be related to protein structure modifications (Fig. 8).

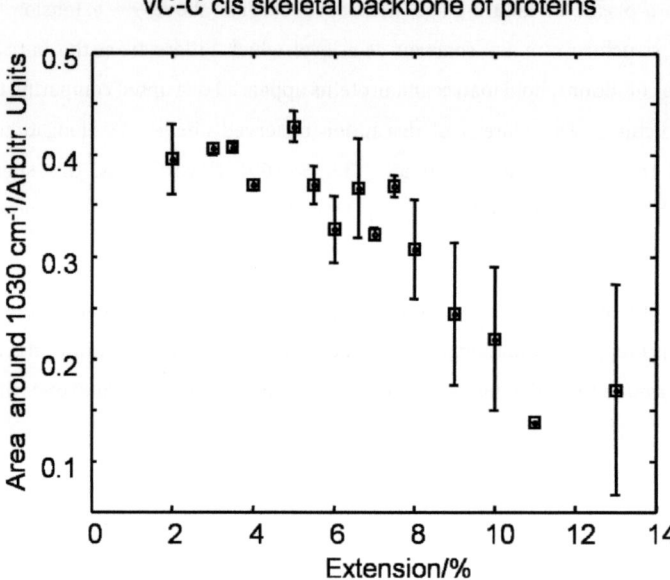

Figure 8: Intensity changes of C-C cis stretching bands of proteins as function of tissue extension.

Regarding the quantity of information provided by Raman spectra, it is thought that taking together all the Raman descriptors would help to characterize mechanical behavior of skin with good precision than using only one peak.

Conclusion

The main objective of this study was to evaluate and substantiate the viscoelastic mechanical behavior of the SC using a noninvasive optical technique and multivariate data analysis. Therefore not only biomechanical skin properties could be assessed in an objective way, but also could allow the detection of subtle changes in SC and consequently in skin morphology and function.

In the present work, we have investigated molecular modifications due to the extension of SC by using Raman spectroscopy. On the one hand, we showed that, when SC is stretched, α helix form of proteins unravels and are converted into β-sheet forms. On the other hand, lipids gauche conformation, which corresponds to less ordered lipid matrix, increases with strain. The present results suggest that the SC strain decreases the skin barrier function.

We develop a partial least square model from which we are able to predict the SC strain status from acquired Raman spectra. We then validate the model by comparing these predictions to the measured elongations using physically measurements of the SC length. The strong linear relationship (R2= 0.990) between the predicted and measured sample elongation demonstrate that there is a correlation between the SC tensile properties and the Raman signature. This result suggests that the conformational modifications of proteins and lipids could provide information about the sample strain status. The methods presented in this study could be a suitable tool that can be extended for clinical applications such as diagnostic of skin discomfort degree and elastic behavior evaluation. In these cases, this tool could be also used to evaluate the effectiveness of dermocosmetic products that are developed to improve the mechanical behavior not only for aesthetic purposes but also to maintain the skin homeostasis conditions.

It was interesting to see that the spectral Raman features correlate well with the extension state of the SC. Further experiments should be performed in vivo and then could be used in the future to develop a model that is able to predict the skin tensile properties. Such model could be extended to different type of measurements as torsion, suction and indentation, hoping that it could give us information of global mechanical behavior of the skin.

Acknowledgements

The authors thank Pr. Pierre Chaminade (Paris Sud University, France) for supervising the PLS data processing.

References

[1] J.L. Lévêque, B. Audoly, Influence of Stratum Corneum on the entire skin mechanical properties, as predicted by a computational skin model, Skin Research and Technology, 19 (2012) 42-46.

[2] M. Geerligs, G.W.M. Peters, P.A.J. Ackermans, C.W.J. Oomens, F.P.T. Baaijens, Does subcutaneous adipose tissue behave as an (anti-)thixotropic material?, Journal of biomechanics, 43 (2010) 1153-1159.

[3] M. Geerligs, G.W.M. Peters, P.A.J. Ackermans, C.W.J. Oomens, F.P.T. Baaijens, Linear viscoelastic behavior of subcutaneous adipose tissue, Biorheology, 45 (2008) 677-688.

[4] G.P. Seehra, F.H. Silver, Viscoelastic properties of acid- and alkaline-treated human dermis: a correlation between total surface charge and elastic modulus, Skin Research and Technology, 12 (2006) 190-198.

[5] D.W. Evans, A.T. Scott, R.D. Teasdall, B.P. Smith, A. Howlett, T. Connor-Kerr, J.L. Sparks, Mechanical properties of lower limb dermis following static and cyclic compression, Biomed Sci Instrum, 48 (2012) 104-111.

[6] C.A. Grant, P.C. Twigg, D.J. Tobin, Static and dynamic nanomechanical properties of human skin tissue using atomic force microscopy: effect of scarring in the upper dermis, Acta Biomater, 8 (2012) 4123-4129.

[7] Y. Mofid, G. Josse, S.n. Gahagnon, A. Delalleau, F.d.r. Ossant, Mechanical skin thinning-to-thickening transition observed in vivo through 2D high frequency elastography, Journal of biomechanics, 43 (2010) 2954-2962.

[8] F.M. Hendriks, D. Brokken, C.W.J. Oomens, D.L. Bader, F.P.T. Baaijens, The relative contributions of different skin layers to the mechanical behavior of human skin in vivo using suction experiments, Medical engineering & physics, 28 (2006) 259-266.

[9] H.V. Tran, F. Charleux, M. Rachik, A. Ehrlacher, M.C. Ho Ba Tho, In vivo characterization of the mechanical properties of human skin derived from MRI and indentation techniques, Comput Methods Biomech Biomed Engin, 10 (2007) 401-407.

[10] M.S. Christensen, C.W. Hargens, III, S. Nacht, E.H. Gans, VISCOELASTIC PROPERTIES OF INTACT HUMAN SKIN: INSTRUMENTATION, HYDRATION EFFECTS, AND THE CONTRIBUTION OF THE STRATUM CORNEUM, J Investig Dermatol, 69 (1977) 282-286.

[11] M. Geerligs, L. van Breemen, G. Peters, P. Ackermans, F. Baaijens, C. Oomens, In vitro indentation to determine the mechanical properties of epidermis, Journal of biomechanics, 44 (2011) 1176-1181.

[12] S. Doubal, P. Klemera, Changes in mechanical properties of skin as a marker of biological age, Sb Lek, 99 (1998) 423-428.

[13] G.E. Pierard, C. Franchimont, C.M. Lapiere, Ageing as shown by the microanatomy and the physical properties of the skin, Int J Cosmet Sci, 2 (1980) 209-214.

[14] T. Hermanns-Le, I. Uhoda, S. Smitz, G.E. Pierard, Skin tensile properties revisited during ageing. Where now, where next?, J Cosmet Dermatol, 3 (2004) 35-40.

[15] Y.S. Papir, K.H. Hsu, R.H. Wildnauer, The mechanical properties of stratum corneum. I. The effect of water and ambient temperature on the tensile properties of newborn rat stratum corneum, Biochim Biophys Acta, 399 (1975) 170-180.

[16] K.S. Wu, W.W. van Osdol, R.H. Dauskardt, Mechanical properties of human stratum corneum: effects of temperature, hydration, and chemical treatment, Biomaterials, 27 (2006) 785-795.

[17] C.S. Nicolopoulos, P.V. Giannoudis, K.D. Glaros, J.C. Barbenel, In vitro study of the failure of skin surface after influence of hydration and preconditioning, Arch Dermatol Res, 290 (1998) 638-640.

[18] M.A. Wolfram, N.F. Wolejsza, K. Laden, Biomechanical properties of delipidized stratum corneum, J Invest Dermatol, 59 (1972) 421-426.

[19] G.K. Menon, G.W. Cleary, M.E. Lane, The structure and function of the stratum corneum, Int J Pharm, 435 (2012) 3-9.

[20] P.M. Elias, Structure and Function of the Stratum Corneum Extracellular Matrix, J Invest Dermatol, 132 (2012) 2131-2133.

[21] Marks, R, The stratum corneum barrier: The final frontier, American Society for Nutrition, Bethesda, MD, ETATS-UNIS, 2004.

[22] C.R. Harding, The stratum corneum: structure and function in health and disease, Dermatologic Therapy, 17 (2004) 6-15.

[23] M. Simon, D. Bernard, A.M. Minondo, C. Camus, F. Fiat, P. Corcuff, R. Schmidt, G. Serre, Persistence of both peripheral and non-peripheral corneodesmosomes in the

upper stratum corneum of winter xerosis skin versus only peripheral in normal skin, J Invest Dermatol, 116 (2001) 23-30.

[24] L. Rodrigues, EEMCO Guidance to the in vivo Assessment of Tensile Functional Properties of the Skin, Skin Pharmacology and Physiology, 14 (2001) 52-67.

[25] G.B. Jemec, J. Serup, Epidermal hydration and skin mechanics. The relationship between electrical capacitance and the mechanical properties of human skin in vivo, Acta dermato-venereologica, 70 (1990) 245-247.

[26] S. Kapoor, S. Saraf, Assessment of viscoelasticity and hydration effect of herbal moisturizers using bioengineering techniques, Pharmacognosy magazine, 6 (2010) 298-304.

[27] A. Delalleau, G. Josse, J.-M. Lagarde, H. Zahouani, J.-M. Bergheau, Characterization of the mechanical properties of skin by inverse analysis combined with the indentation test, Journal of biomechanics, 39 (2006) 1603-1610.

[28] B.F. Van Duzee, The influence of water content, chemical treatment and temperature on the rheological properties of stratum corneum, J Invest Dermatol, 71 (1978) 140-144.

[29] R.H. Wildnauer, J.W. Bothwell, Douglass, Stratum corneum biomechanical properties. I. Influence of relative humidity on normal and extracted human stratum corneum, J Invest Dermatol, 56 (1971) 72-78.

[30] A. Lundstrom, G. Serre, M. Haftek, T. Egelrud, Evidence for a role of corneodesmosin, a protein which may serve to modify desmosomes during cornification, in stratum corneum cell cohesion and desquamation, Arch Dermatol Res, 286 (1994) 369-375.

[31] A.V. Rawlings, I.R. Scott, C.R. Harding, P.A. Bowser, Stratum corneum moisturization at the molecular level, J Invest Dermatol, 103 (1994) 731-741.

[32] Rawlings, V. A, Watkinson, A, Harding, R. C, Ackerman, C, Banks, J, Hope, J, Scott, R. I, Changes in stratum corneum lipid and desmosome structure together with water barrier function during mechanical stress, Society of Cosmetic Chemists, New York, NY, ETATS-UNIS, 1995.

[33] G.L. Wilkes, I.A. Brown, R.H. Wildnauer, The biomechanical properties of skin, CRC Crit Rev Bioeng, 1 (1973) 453-495.

[34] Y. Tomita, M. Akiyama, H. Shimizu, Stratum corneum hydration and flexibility are useful parameters to indicate clinical severity of congenital ichthyosis, Exp Dermatol, 14 (2005) 619-624.

[35] G.S.K. Pilgram, D.C.J. Vissers, H. van der Meulen, S. Pavel, S.P.M. Lavrijsen, J.A. Bouwstra, H.K. Koerten, Aberrant Lipid Organization in Stratum Corneum of Patients with Atopic Dermatitis and Lamellar Ichthyosis, 117 (2001) 710-717.

[36] T.R. Tilleman, M.M. Tilleman, M.H. Neumann, The elastic properties of cancerous skin: Poisson's ratio and Young's modulus, Isr Med Assoc J, 6 (2004) 753-755.

[37] S. Wold, M. Sjostrom, L. Eriksson, PLS-regression: a basic tool of chemometrics, Chemometrics and Intelligent Laboratory Systems, 58 (2001) 109-130.

[38] R. Bro, Multiway calibration. Multilinear PLS, Journal of Chemometrics, 10 (1996) 47-61.

[39] I. Barman, C.R. Kong, N.C. Dingari, R.R. Dasari, M.S. Feld, Development of robust calibration models using support vector machines for spectroscopic monitoring of blood glucose, Anal Chem, 82 (2010) 9719-9726.

[40] M. Nicollier, P. Agache, J.L. Kienzler, R. Laurent, R. Gibey, N. Cardot, J.C. Henry, Action of trypsin on human plantar stratum corneum. An ultrastructural study, Arch Dermatol Res, 268 (1980) 53-64.

[41] L.C. Morejohn, J.N. Pratley, Differential effects of trypsin on the epidermis of Rana catesbeiana. Observations on differentiating junctions and cytoskeletons, Cell Tissue Res, 198 (1979) 349-362.

[42] V.S. Thakoersing, M.O. Danso, A. Mulder, G. Gooris, A. El Ghalbzouri, J.A. Bouwstra, Nature versus nurture: does human skin maintain its stratum corneum lipid properties in vitro?, Experimental Dermatology, 21 (2012) 865-870.

[43] M. Förster, M.A. Bolzinger, M.R. Rovere, O. Damour, G. Montagnac, S. Briançon, Confocal Raman microspectroscopy for evaluating the stratum corneum removal by 3 standard methods, Skin Pharmacol Physiol, 24 (2011) 103-112.

[44] K. Biniek, K. Levi, R.H. Dauskardt, Solar UV radiation reduces the barrier function of human skin, Proceedings of the National Academy of Sciences, 109 (2012) 17111-17116.

[45] S. Tfaili, C. Gobinet, G. Josse, J.-F. Angiboust, M. Manfait, O. Piot, Confocal Raman microspectroscopy for skin characterization: a comparative study between human skin and pig skin, Analyst, 137 (2012) 3673-3682.

[46] J.L. Bruneel, J.C. Lassègues, C. Sourisseau, In-depth analyses by confocal Raman microspectrometry: experimental features and modeling of the refraction effects, Journal of Raman Spectroscopy, 33 (2002) 815-828.

[47] A. Tfayli, O. Piot, M. Manfait, Confocal Raman microspectroscopy on excised human skin: uncertainties in depth profiling and mathematical correction applied to dermatological drug permeation, J Biophotonics, 1 (2008) 140-153.

[48] A. Savitzky, M.J.E. Golay, Smoothing and differentiation of data by simplified least squares procedures, Anal. Chem., 36 (1964) 1627-1639.

[49] J. Zhao, H. Lui, D.I. McLean, H. Zeng, Automated autofluorescence background subtraction algorithm for biomedical Raman spectroscopy, Appl Spectrosc, 61 (2007) 1225-1232.

[50] S. Wold, H. Antti, F. Lindgren, J. Ã–hman, Orthogonal signal correction of near-infrared spectra, Chemometrics and Intelligent Laboratory Systems, 44 (1998) 175-185.

[51] H.T. Eastment, W.J. Krzanowski, Cross-Validatory Choice of the Number of Components From a Principal Component Analysis, Technometrics, 24 (1982) 73-77.

[52] G. Diana, C. Tommasi, Cross-validation methods in principal component analysis: A comparison, Statistical Methods and Applications, 11 (2002) 71-82.

[53] P.J. Caspers, G.W. Lucassen, R. Wolthuis, H.A. Bruining, G.J. Puppels, In vitro and in vivo Raman spectroscopy of human skin, Biospectroscopy, 4 (1998) S31-39.

[54] A. Tfayli, E. Guillard, M. Manfait, A. Baillet-Guffroy, Thermal dependence of Raman descriptors of ceramides. Part I: effect of double bonds in hydrocarbon chains, Anal Bioanal Chem, 397 (2010) 1281-1296.

[55] E. Guillard, A. Tfayli, M. Manfait, A. Baillet-Guffroy, Thermal dependence of Raman descriptors of ceramides. Part II: effect of chains lengths and head group structures, Anal Bioanal Chem, 399 1201-1213.

[56] S. Wartewig, R. Neubert, W. Rettig, K. Hesse, Structure of stratum corneum lipids characterized by FT-Raman spectroscopy and DSC. IV. Mixtures of ceramides and oleic acid, Chemistry and Physics of Lipids, 91 (1998) 145-152.

[57] A. Tfayli, E. Guillard, M. Manfait, A. Baillet-Guffroy, Raman spectroscopy: feasibility of in vivo survey of stratum corneum lipids, effect of natural aging, Eur J Dermatol.

[58] Shetty, G, Kendall, C, Shepherd, N, Stone, N, Barr, H, Raman spectroscopy : elucidation of biochemical changes in carcinogenesis of oesophagus, Nature Publishing Group, Basingstoke, ROYAUME-UNI, 2006.

[59] Chan, W. James, Taylor, S. Douglas, Zwerdling, Theodore, Lane, M. Stephen, Lhara, Ko, Huser, Thomas, Micro-Raman spectroscopy detects individual neoplastic and normal hematopoietic cells, Cell Press, Cambridge, MA, ETATS-UNIS, 2005.

[60] N. Stone, C. Kendall, J. Smith, P. Crow, H. Barr, Raman spectroscopy for identification of epithelial cancers, Faraday Discuss, 126 (2004) 141-157; discussion 169-183.

[61] S. Farquharson, C. Shende, F.E. Inscore, P. Maksymiuk, A. Gift, Analysis of 5-fluorouracil in saliva using surface-enhanced Raman spectroscopy, Journal of Raman Spectroscopy, 36 (2005) 208-212.

[62] A.T. Tu, Raman Spectroscopy in Biology: Principles & Applications, Books on Demand, 1982.

[63] Biswas, Nilanjana, Waring, J. Alan, Walther, J. Frans, Dluhy, A. Richard, Structure and conformation of the disulfide bond in dimeric lung surfactant peptides SP-B[1-] [25] and SP-B[8-] [25], Elsevier, Amsterdam, PAYS-BAS, 2007.

[64] B.E. Weiss-Lopez, M.H. Goodrow, W.K. Musker, C.P. Nash, Conformational dependence of the disulfide stretching frequency in cyclic model compounds, Journal of the American Chemical Society, 108 (1986) 1271-1274.

[65] M. Gniadecka, O. Faurskov Nielsen, D.H. Christensen, H.C. Wulf, Structure of water, proteins, and lipids in intact human skin, hair, and nail, J Invest Dermatol, 110 (1998) 393-398.

[66] I. Notingher, C. Green, C. Dyer, E. Perkins, N. Hopkins, C. Lindsay, L.L. Hench, Discrimination between ricin and sulphur mustard toxicity in vitro using Raman spectroscopy, J R Soc Interface, 1 (2004) 79-90.

[67] E. Katainen, M. Elomaa, U.-M. Laakkonen, E. Sippola, P. Niemelä, J. Suhonen, K. Järvinen, Quantification of the Amphetamine Content in Seized Street Samples by Raman Spectroscopy, Journal of Forensic Sciences, 52 (2007) 88-92.

[68] A.C. PARK, C.B. BADDIEL, Rheology of stratum corneum-I: A molecular interpretation of the stress-strain curve, J. Soc. Cosmet. Chem., 23 (1972) 3-12.

[69] Y. Yuan, R. Verma, Measuring microelastic properties of stratum corneum, Colloids and Surfaces B: Biointerfaces, 48 (2006) 6-12.

[70] J. Conroy, A.G. Ryder, M.N. Leger, K. Hennessey, M.G. Madden, Qualitative and quantitative analysis of chlorinated solvents using Raman spectroscopy and machine learning, (2005) 131-142.

[71] H. Schulz, M. Baranska, Identification and quantification of valuable plant substances by IR and Raman spectroscopy, Vibrational Spectroscopy, 43 (2007) 13-25.

[72] C. Krafft, L. Neudert, T. Simat, R. Salzer, Near infrared Raman spectra of human brain lipids, Spectrochimica Acta Part A: Molecular and Biomolecular Spectroscopy, 61 (2005) 1529-1535.

[73] E. Ó Faoláin, M.B. Hunter, J.M. Byrne, P. Kelehan, M. McNamara, H.J. Byrne, F.M. Lyng, A study examining the effects of tissue processing on human tissue sections using vibrational spectroscopy, Vibrational Spectroscopy, 38 (2005) 121-127.

CHAP VII. EFFETS DES MOLECULES HYDRATANTES SUR LA STRUCTURE DU *STRATUM CORNEUM* : SPECTROSCOPIE VIBRATIONNELLES ET CLS

VII.1.   PREAMBULE

D'un point de vue dermatologique, un traitement hydratant a plusieurs objectifs, réparer les fines fissures de la peau, augmenter sa teneur en eau, réduire la perte d'eau à travers l'épiderme et ainsi restituer la fonction barrière de la peau [172-174]. L'interaction des hydratants avec la peau se traduit phénotypiquement par une peau lisse et souple.

Outre les aspects physiques qu'offre l'hydratation, la quantité d'eau dans le SC est un facteur déterminant dans l'activité enzymatique qui, à son tour, contrôle la quantité des NMF, des lipides ainsi que la maturation des cornéocytes [175, 176]. Ces composantes contrôlent à leur tour la quantité d'eau dans le SC. En considérant ce cycle, Il est évident que la restauration de l'hydratation par apport d'eau est loin d'être un simple traitement symptomatique, mais aussi attaque le cœur du problème de la « peau sèche ». Ce traitement peut rétablir les réactions biochimiques qui garantissent un bon fonctionnement du SC.

Les hydratants peuvent être classés en quatre groupes : (a) les occlusifs qui sont de nature grasse, empêchent la perte de l'eau ; (b) les promoteurs de la compacité qui, en interagissant avec la matrice lipidique du SC, augmentent la barrière à la perte d'eau, (c) les humectants, des substances hygroscopiques, qui captent l'eau; (d) les émollients qui rendent la peau plus lisse et plus douce. Comme les émollients sont définis sur un critère qui ne fait pas référence à l'eau, ils peuvent également avoir une conjugaison de propriétés occlusives et hydrophiles.

Dans cette thèse, nous mettons en œuvre les méthodes spectroscopiques Raman et infrarouge dans la détermination des mécanismes d'action d'un certain nombre de molécules hydratantes en décrivant l'influence de ces molécules sur l'absorption de l'eau et sur la fonction barrière du SC.

ARTICLE 4 : SPECTROSCOPIE VIBRATIONNELLES COUPLE A L'ANALYSE CLASSIQUE
PAR MOINDRES CARRE : UNE NOUVELLE APPROCHE POUR LA DETERMINATION DES
MECANISMES D'ACTION D'AGENTS HYDRATANTS CUTANES

Contexte de l'étude: La sécheresse cutanée est un symptôme omniprésent dans divers types de maladies de la peau. De ce fait, un grand nombre de produits de soin de la peau est développée en particulier à des fins d'accompagnement thérapeutique aux pathologies cutanées associant un état xérotique. L'analyse de l'hydratation du SC et de sa fonction barrière est essentielle pour évaluer les effets des agents hydratants.

Méthodes: Dans ce travail, les spectroscopies Raman confocale et ATR-FTIR ont été utilisées pour étudier, au niveau moléculaire, les effets des différents agents hydratants sur l'état de la barrière du SC et du degré d'hydratation. Les différents composants chimiques ont été appliqués localement sur des SC hydratés: des acides, des alcools, des esters, des triglycérides, des dérivés d'acides gras et un mélange de lipides à longue chaîne.

Tout d'abord, les propriétés hygroscopiques des différents composants ont été déterminées par leur capacité à fixer l'eau par l'analyse des échantillons de SC maintenus dans une atmosphère à 90% d'HR. Ensuite, la fonction barrière à la perte d'eau a été évaluée après le processus de déshydratation (4% d'HR) par spectroscopie Raman. Dans ce but, l'analyse par moindres carrés classiques (CLS) a permis d'extraire les signatures Raman du SC à partir du signal brut contenant le mélange des signaux du SC et de la molécule hydratante appliquée. Ceci conduit à la mise en évidence les changements supramoléculaires au niveau du SC. La modification ultra-structurale du SC pourrait expliquer l'effet réorganisateur de certains hydratants sur la matrice lipidique.

De plus, la cinétique de pénétration des molécules à travers le SC a été étudiée pendant 2 heures. Le manque de pénétration et la diminution des pertes d'eau déterminent la propriété occlusive. Les émollients n'ont pas d'effet sur l'hydratation.

Résultats: En utilisant cette approche methodologique, le glycérol et le propylène glycol ont été caractérisés comme humectants; la lanoline a montré une action occlusive, l'acide lactique est à la fois humectant et promoteur de la barrière et l'éthyl hexyle palmitate et le triglycéride d'acide caprylique / caprique semblaient être des émollients.

Ces observations sont en accord avec la littérature sur la base de la classification actuelle fondée sur des mesures biométriques.

Conclusion: La méthode développée caractérise de façon non invasive le mécanisme d'action des molécules testées. Cette approche spectroscopique peut améliorer les connaissances sur les relations structure-activité des nouvelles molécules et contribuer à la démonstration de la preuve de concept thérapeutique efficace face aux divers dysfonctionnements de la peau.

ARTICLE 4

Skin Research and Technology 2013; 0: 1–11
Printed in Singapore · All rights reserved
doi: 10.1111/srt.12117

© 2013 John Wiley & Sons A/S.
Published by John Wiley & Sons Ltd

Skin Research and Technology

Vibrational spectroscopy coupled to classical least square analysis, a new approach for determination of skin moisturizing agents' mechanisms

R. Vyumvuhore[1], A. Tfayli[1], M. Manfait[2] and A. Baillet-Guffroy[1]

Vibrational spectroscopy coupled to classical least square analysis, a new approach for determination of skin moisturizing agents' mechanisms

Raoul Vyumvuhore[a], Ali Tfayli *[a], Michel Manfait[b], Arlette Baillet-Guffroy[a].

[a]Laboratory of Analytical Chemistry, Analytical Chemistry Group of Paris-Sud (GCAPS-EA4041), Faculty of Pharmacy, University Paris Sud , Chatenay-Malabry, France

[b]MéDIAN Unit, CNRS UMR 6237, Faculty of Pharmacy, Univ. Reims Champagne Ardennes, Reims, France

Abstract:

Background: Skin dryness is an omnipresent symptom in various types of skin disorders. Thereby, a large panel of skin care products is developed for therapeutic purposes. However, still a lack of non-invasive methods to determine the mechanisms of action of moisturizers, at the molecular level.

Methods: In the present study, confocal Raman spectroscopy coupled to classical least square analyses and ATR-FTIR were used to investigate, the effect of different molecules on the *stratum corneum* (SC) hydration degree and barrier state. Firstly, water caption capacity was determined by analyzing samples at 90% RH, secondary the water barrier function was evaluated after the dehydration process (4 % RH). The molecules penetration kinetics across SC was also studied during 2 hours.

Results: Using the present approach **glycerin and propylene glycol** were found to be **humectants**; **lanoline** showed **occlusive** action, **lactic acid** has both **humectant and barrier enhancer**, properties and **ethylhexyl palmitate and caprylic/capric acid triglyceride** seemed to be **emollients**. Those observations are in accordance with literature.

Conclusion: The present method non-invasively characterizes the mechanism of action of tested molecules. This may improve knowledge on new molecules structure-activity relationship and help to make an effective therapeutic concept dealing with the various skin dysfunctions.

Keywords: Raman spectroscopy, ATR-FTIR, classical least square analysis, *stratum corneum* barrier function, skin moisturizers

Introduction

Maintaining adequate water content of the skin and protecting its barrier function are essential for healthy and youthful-looking skin. Environmental factors such as dry climates, exposure to wind, cold weather and skin diseases are common causes of dry skin [1-3]. The most affected part of the skin is the outermost layer of the epidermis, the *Stratum Corneum* (SC). It is composed of two major physiological entities: the corneocytes with their natural moisturizing factors (NMF) content and an intercellular lipid bilayer matrix [4]. Proper function of both components ensures skin integrity and good hydration [5].

In normal physiological conditions, the skin minimizes excessive water loss by regulating the production of intercellular skin lipids and NMF. Meanwhile, at extreme physiological and pathological conditions the effectiveness of this regulating process is limited. The use of moisturizing treatments is thereby needed.

Moisturizers alleviate skin dryness by increasing water content (humectants) or reducing water loss (occlusive oils and lipid modulating agents).

Humectants are hygroscopic agents. In short term, they attract water into the SC and increase hydration level [6, 7]. However, for most of them, in longer terms water evaporates from the surface [8].

Occlusive agents are oily substances [9]. Topically applied, they reduce transepidermal water loss (TEWL). They are thought to form an inert film on the skin surface and thus impair evaporation of skin moisture. However, recent findings indicate that some molecules may also diffuse deeper into the SC and modify endogenous epidermal lipids and the rate of barrier recovery loss [10, 11].

Lipid modulating or "barrier repairing" agents are less described in the literature and their chemical natures are not yet established. They act directly on the lipid barrier by improving abnormal function or preventing deterioration of the normal state. Those substances have shown their ability to reduce the expression of skin dryness pathologies like atopic eczema [12].

The analysis of the skin barrier properties is important to evaluate the effects of topically applied substances. Standard methods for characterizing epidermal barrier function like TEWL can be affected by different factors, like water content of the applied formulation and room temperature. In their study, Vilaplana et al. applied an ammonium lactate based moisturizer on atopic skin; no effect on TEWL was detected although clinical appearance was improved [13]. Furthermore, excessive hydration of the SC layer may also impair its diffusion resistance by creating interfacial defects in the lipid bilayer. Thus, an increased TEWL may reflect a decreased as well as an increased level of hydration [14]. Moreover, no information at the molecular level can be obtained using this method.

Alternative methods such Laser scanning microscopy have proved their effectiveness for the characterization of the skin barrier structure [15]. The use of confocal Raman spectroscopy allowed noninvasively monitoring the skin hydration, molecular distribution and structure [16].

The aim of the present study was to access the relationship between the chemical structure of different moisturizers and their mode of action. This will be defined by their effect on the hydration level and on the SC barrier state. Different moisturizing agents with different chemical structure were then used: acids, alcohols, esters, triglycerides, fatty acid derivative and a mix of long chain lipids.

Raman analyses were performed to evaluate their effect on water content [17-20], protein structure [21-23] and barrier function [24-28]. Classical least square analyses (CLS) allowed to extract Raman signatures of the moisturizers and to highlight molecular changes on the SC. Then, penetration kinetics of moisturizers across SC were studied using ATR-FTIR spectroscopy in order to evaluate whether the molecules act in deeper layers or as occlusive.

Referring to our results, the investigation of the structure-activity relationship of dermo-cosmetic ingredients could guide new products development and help to establish effective products in dermatological field.

Materials and methods

Isolation of *stratum corneum*

Human abdominal skin was obtained after plastic surgeries from 5 different female patients aged between 45 and 52 and stored at -20°C. SC samples were prepared using enzymatic digestion method [29-33]. The subcutaneous fat and connective tissues were first removed. Then, the skin tissue was immersed in distilled water (Milli-Q reagent water system) at 60°C for 1 minute. The epidermis was subsequently removed from the dermis. The epidermis was then placed, SC side up, onto a filter paper imbibed with a 0.2% trypsin solution (0.2% in distilled water; Sigma-T4665) for 1 hour. The SC was carefully separated from the underlying epidermis with tweezers. The piece of SC was spread in a water container at ambient temperature and was settled onto a greaseproof paper. The isolated SC was dried in desiccators (containing P_2O_5).

Moisturizers

Lanolin (LAN) was obtained from Alfa Aesar GmbH (Karlsruhe, Germany), **glycerol(GLY)**and **propylene glycol (PRO)** were purchased from Merck (Schurchart, Germany), **ethylhexyl palmitate (PAL)** was provided by Chemos GmbH (Regenstauf, Germany), **lactic acid (LAC)** was obtained from Carl Roth GmbH (Karlsruhe, Germany) and **caprylic/capric acid triglyceride (CAP)** from Evonik Industries AG Goldschmidtstrasse (Essen, Germany).

Humidity controlled environment

A humidity controlled chamber, based on an airflow principle, was built around the microscope [16]. Air flow was controlled by volume flow controllers (Dynaval Air, Air Liquide, Paris, France). A hygrometer/thermometer (with ± 2% accuracy) was used to measure relative humidity. To obtain the desired humidity level, dry and humid air was mixed in an appropriate ratio. The temperature of the air conditioned room was maintained at 20°C±1°C.

Methodology

- Raman spectroscopy investigation: Intact SC samples were adhered on CaF$_2$ slides (outer surface up) and were placed in a humidity controlled chamber under the microscope interfaced to the Raman micro spectrometer. First, samples were allowed to equilibrate for 4 hours at 90% RH and 21°C. Moisturizers were then topically applied (2µl/cm^2) and left to act for additional 2 hours at 90% RH. Raman spectra were then acquired by focusing on the surface of the SC (8µm spatial resolution). After that, RH was set to 4% and 4 hours later Raman spectra were acquired once again.

- Classical least square analyses (CLS) allowed to extract Raman signatures of the SC from the mixture and to highlight molecular changes on the SC including water content information.

- The molecules penetration kinetics across SC was also studied using ATR-FTIR spectroscopy (during 2 hours).

The modifications in SC water content and lipids structure as well as the penetration profile allowed us to conclude on humectant, occlusive and lipid modulating properties of the applied molecules (Figure 1).

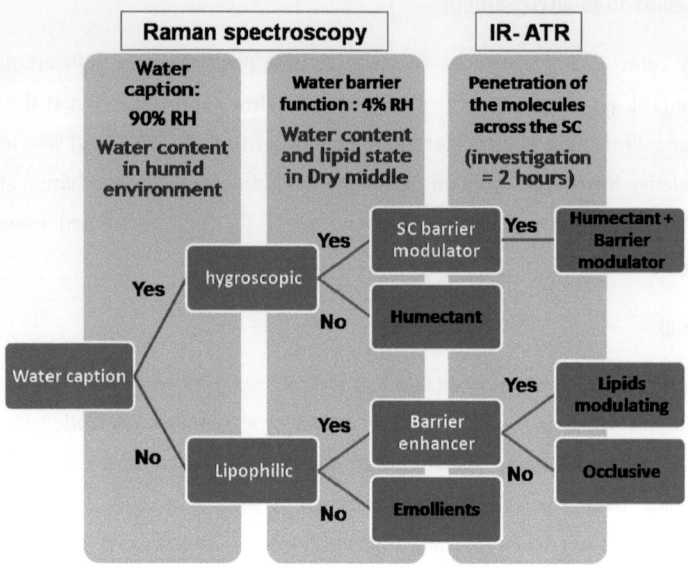

Figure 1: Methodology determining the moisturizing mechanisms of topically applied molecules

ATR-FTIR spectroscopy

ATR-FTIR spectra were obtained using a Spectrum 2000 spectrophotometer PERKIN ELMER (Courtaboeuf, France) equipped with an ATR attenuated total reflection diamond crystal (Golden Gate – Specac) and a liquid-nitrogen cooled, mercury cadmium telluride (MCT) detector. ATR-FTIR spectra were recorded using spectrum software (PERKIN ELMER) in the 550-4000 cm^{-1} range with 4 cm^{-1} resolution and 16 accumulations per spectrum.

To study the penetration kinetics across the SC, hydrated SC samples, sized 1 cm^2, were adhered on the 1 mm^2 surface of the ATR crystal and allowed to dry for 24 hours. Moisturizers were topically applied on the middle part of SC (2µl/cm^2) followed by infrared spectra acquisition every 2 minutes during 2 hours. The applied molecules have specific peaks different from those of the SC; this allowed us to follow the penetration of the molecules. Changes in the collected spectra appeared when the molecules reached the lower side of the SC. Characteristic spectral features of the used molecules were detected by subtracting each spectrum of the treated samples by molecules from the SC spectrum before treatment. The first detection of these specific peaks was considered as the precise penetration time of the molecule. This experience was repeated three times and same results were observed.

Raman spectroscopy analysis

A Horiba JobinYvon LabRAM Raman microspectrometer (Horiba Scientific, Lille, France) was used for data acquisition. A video image of the sample was used for accurate positioning of the laser spot on the sample. A 660 nm double pumped Nd:YLF laser (Toptica Photonics, Munich, Germany) with 10 mW power, on the sample, was used. The 660 nm excitation wavelength was chosen because it gives a weak background in the fingerprint region [34] and a high Raman Stokes signal in the high wavenumbers region. A long focal microscope objective PL Fluotar L 100X/ NA 0.75 WD 4.7 (Leica, Mannheim, Germany) was used to focus the laser light on the sample and to collect the back diffused light. The confocal hole was set at 150 µm for all measurements. The in-depth spot size was around 8 µm allowing collecting signal from the half of the sample thickness. The collected light was filtered through a notch filter and dispersed with a 4 cm^{-1} spectral resolution using a holographic grating of 950 grooves/mm. The Raman Stokes signal was recorded with a Charge-Coupled Device detector: CCD camera (Andor technology, Belfast, UK) containing 1024 x 256 pixels. Spectral acquisition was performed using Labspec 5 software (Horiba Scientific, Lille, France).

Raman Signal processing

All spectra were smoothed using Savitsky Golay algorithm with 11 points and 2nd polynomial degree [35]. The baseline was corrected using an automatic polynomial function [36]. Normalization of the spectra was done using unit vector method.

Classical least square (CLS) analysis was carried out using in-house developed software that operates in the Matlab environment (The MathWorks, Inc., Natick, Massachusetts, USA). The CLS method is used to determine the contribution of each compound in a mixture. This method is effective when the spectrum of a mixture is a linear combination of the spectra of individual components of the mixture multiplied by the concentration of those components [37]. In the present case, the contribution of 2 spectra: X1 and X2 in a mixture spectrum need to be determined. Their respective concentrations were A1 and A2. The mixture spectrum (Y) can thus be described by the following equation:

$Y_i = A1_i * X1 + A2_i * X2 + \varepsilon_i$;   Equation 1

Where ε_i includes noise and spectral variations.

The measured spectra of the mixture and the concentration values A1 and A2 are adjusted until least squares fit is produced, thus give the estimated concentration of each individual component of the mixture.

For each Raman descriptor, a student t-test was used to determine if untreated sample is significantly different from treated samples. Data were presented as mean ± standard deviation. Since a physiological state of the SC is characterized by many Raman features, the interpretation about a structure modification was based at least on 2 descriptors.

Results

Raman measurements and classical least square (CLS) analyses

Raman spectroscopy, coupled with signal extraction data processing, was used to evaluate the effect of topically applied products on skin structure.

Validation of the CLS method

For validation, CLS was first performed on spectra of different GLY:PRO mixtures. Proportions 1:1, 1:2, 1:3, 1:4 and 1:5 were used.

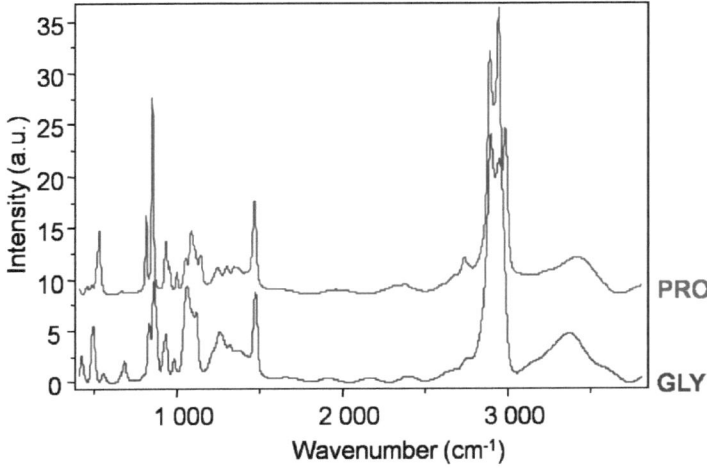

Figure 2: FTIR Spectra of glycerol and propylene glycol

Despite the spectral similarity and numerous overlapping bands between the spectra of the two components (Fig. 2) CLS models (equation 1) gave good estimations of the molecules' proportions (table 1).

Real proportion (GLY: PRO)	1 : 1,0	1 : 2,0	1:3,0	1 : 4,0	1 : 5,0
CLS calculated proportion (GLY: PRO)	1 : 1,2	1 : 2,1	1 : 3.2	1 : 4,2	1 : 4,8

Table 1: Mixtures of Glycerin and Propylene glycol; Comparison between real proportions and CLS calculated proportions

Determination of changes in SC structure

For accurate analysis of SC structure modifications, the moisturizer Raman signal should be removed from the measured spectrum of treated sample. To extract SC spectra from samples treated with moisturizers, each spectrum was individually fitted as described previously in equation 1 ($Y_i = A1_i*X1 + A2_i*X2 + \varepsilon_i$). X1 was the moisturizer reference spectrum and X2 the mean spectrum of untreated SC. ε_i thus contained residual fitting error (variability between different points of untreated SC) and the variations due to the interaction between the moisturizing agent and the SC.

Based on Equation 1, the spectral contribution of the moisturizers ($A1_i*X1$) can be removed from the measured spectrum (Y_i) (equation 2).

$$E_i = Y_i - A1_i*X1 \qquad \text{Equation 2}$$

A confidence interval (Fig. 3A) was determined from the different ε_i of the all individual untreated SC spectra.

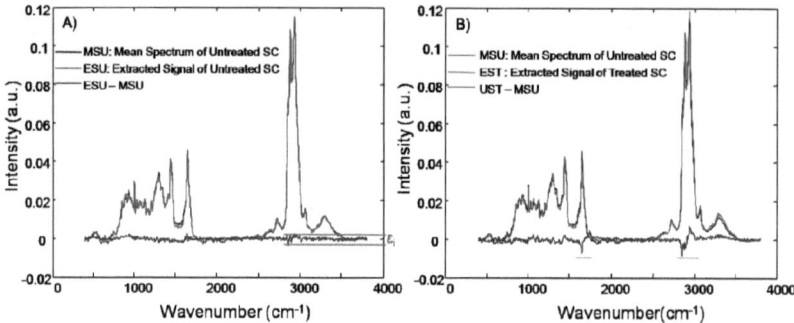

Figure 3: Determination of the peaks modified by the treatment: A) ESU – MSU = Difference between mean spectrum of untreated SC and extracted individual spectrum of untreated SC; B) EST – MSU = Difference between mean spectrum of untreated SC and extracted spectrum of treated SC. εi = confidence interval

When the present signal extraction is applied on treated SC spectra, spectral variations higher than confidence interval were considered as due to the interaction with the moisturizer (Fig. 3B). Thus, the bands affected by the different moisturizers were indentified and investigated in details.

Moisturizers effects on SC water content

Firstly, water content was measured on the SC surface maintained at 90% RH. The olefin (=C-H) vibration band is a SC specific band which is not found in the spectra of the tested molecules. Its relative intensity seems to be constant compared to the classically used band for normalization; νC-H (2910-2955 cm^{-1}). Therefore, OH band areas were divided by the area between 3040 and 3096 cm^{-1} to compensate for the difference in signal intensity.

Total water content of the SC was calculated from the area under curve (AUC) between 3100 and 3620 cm^{-1} (Fig. 4) [16]. It should be noted here that, first Raman measurements were acquired 2 h after product application. Interestingly, the molecules GLY, LAC and PRO appear to increase the water levels in the SC (Fig. 4A). Then, we can say that these moisturizers attract water from the surroundings or favor conformation changes giving to the SC a more hygroscopic property. They are humectants. For samples treated with moisturizer LAN and PAL the water levels are lower in the first superficial 8µm of the sample , this is most probably caused by action of applied products which might push water molecules from the superficial towards the depth of SC.

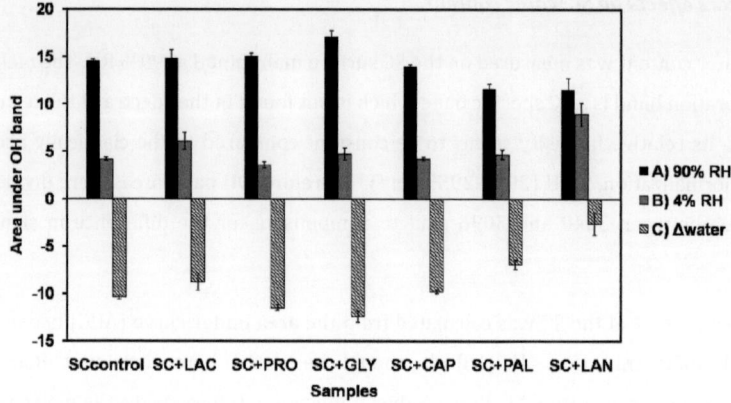

Figure 4: Mean value of 5 measurements of global water content for the different samples: band area in the range of 3100-3620 cm^{-1}. A) water caption of sc samples 2 hours after topical application of the molecules: rh = 90%; b) residual water content after drying process: 4 hours at rh = 4%; c) water loss from 90% to 4% rh of samples treated with different molecules: difference between oh band at 4% and 90% rh. All spectra were normalized on the band area (3040-3096 cm^{-1}).

Secondary, the residual water content in SC was measured after 4 hours in dry environment (RH=4%). From this quantification, LAN appeared to markedly increase the water retention capacity of the SC (Fig. 4B). In second place comes LAC. The mechanism of action of this product is defined by its penetration in the SC and interaction with the SC components. This will be illustrated and discussed later in the manuscript. Other molecules seemed to have no effect on water retention capacity of the SC.

Finally, the water loss was evaluated by calculating the difference between water content at 4% and 90% RH. For illustration, water loss value from each treatment is given in Fig. 4C. The samples treated with molecules GLY and PRO lost a big amount of their water content, while the water loss of the samples treated with LAN, LAC and PAL appeared to be small compared to untreated samples. The latter could be effective molecules to prevent skin dehydration in dry environment.

Since different types of water are present in the SC (namely bound and unbound water), it is interesting to investigate their distribution in different samples after the drying process.

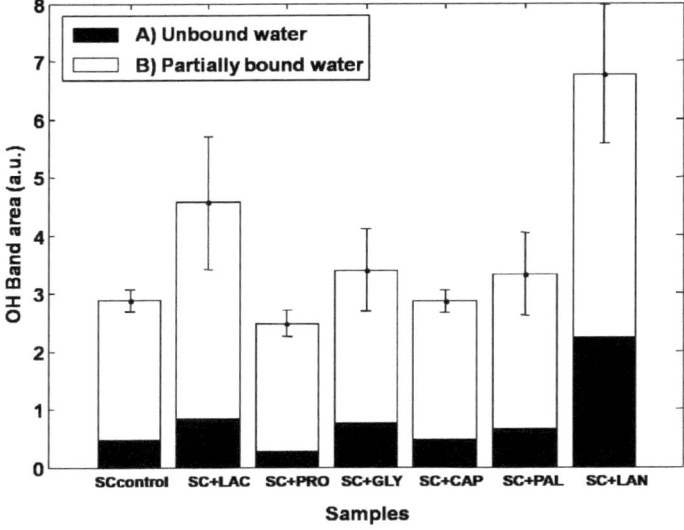

Fig. 5 illustrates the remaining water content in SC after 4hours drying at 4% of RH. Unbound water was obtained from AUC between 3420 and 3620 cm⁻¹ while partially bound water was evaluated from AUC of the band centered at 3300 cm⁻¹ (3245-3400 cm⁻¹) [16].

Except for samples treated with LAN (Fig. 5A), almost no unbound water content remained after 4 hours in a dry environment (4 %), while residual partially bound water was found to be higher for samples treated with LAC and LAN (Fig. 5B).

The LAN capacity to hold both unbound and bound water could be associated to an occlusive action preventing water loss from the SC. On the other hand, the action of LAC was limited to the retention of bound water; this could be associated with modifications in protein structure and lipids barrier.

For a better understanding of the action mode of the tested molecules on water retention capacity, SC protein structure and lipid barrier state were evaluated after the drying process.

Moisturizers effects on stratum corneum proteins' secondary structure

Interaction between moisturizing agents and keratin fibers may modify the keratin structure. In order to elucidate the effect of the moisturizers on the SC protein secondary structure, we examined variation in the β sheet/ α-helix ratio. This information was obtained using features from Amide I region (1600-1700 cm^{-1}) and Amide III band (1190-1370 cm^{-1}). The α helix form of the keratin band arise around 1653 cm^{-1} and 1279 cm^{-1}, while the β sheets is around 1672 cm^{-1} and 1249 cm^{-1} [21-23].

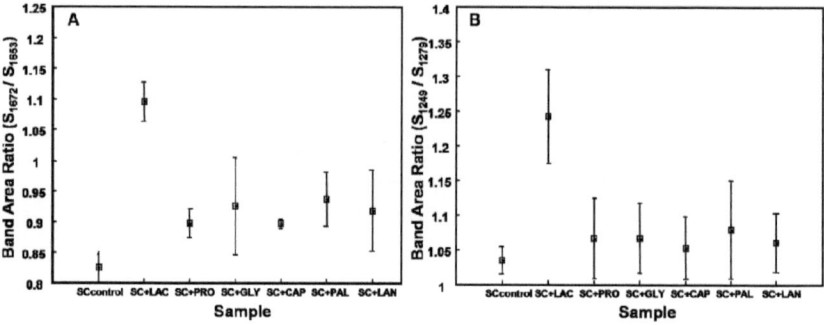

Figure 6: Secondary structure of sc protein as function of applied moisturizers after drying process (rh = 4%). The conformational changes were estimated using β-sheet/α-helix ratio of: a) ratio of amide i sub-bands (s_{1672} /s_{1653}); b) ratio of amide iii sub-bands (s_{1249} /s_{1279})

The present results showed that the protein secondary structure was modified by the applied moisturizers (Fig. 6). Particularly, sample treated with LAC was characterized with a greater conversion of a fraction of α-helixes into β-sheets. One can assume that the structural changes of SC protein may be a key point in understanding the moisturizing mechanism of topically applied molecules.

It's thought that the increase in the water holding capacity of SC treated with LAC is caused by changes in the interaction between water molecules and the SC protein. The LAC may modify the SC protein structure and making available protein site to interact with water molecules. This is in accordance with previous studies suggesting that lactate salts interact with SC components and increase their bound water content [38]. Therefore, the water holding capacity of the SC may be regulated not only by the NMF content but also by the structure of the SC protein. Indeed, It has been shown that keratin gene mutations, which affect the structure of keratin, cause dry skin symptom [39].

Moisturizers effects on stratum corneum lipid's conformational order

The lipid conformational order in the SC was evaluated using the νC–C and the νC–H vibration modes of lipids hydrocarbon chain. The *trans/gauche* conformers ratio were analyzed by calculating I_{1060}/I_{1080} ratio and The $\nu_{asym}CH_2$ (2885 cm^{-1})/$\nu_{sym}CH_2$ (2850 cm^{-1}) ratio. High values of these ratios are associated with a compact state in the lipid packing while a decrease is indicative of a loosening [24-28].

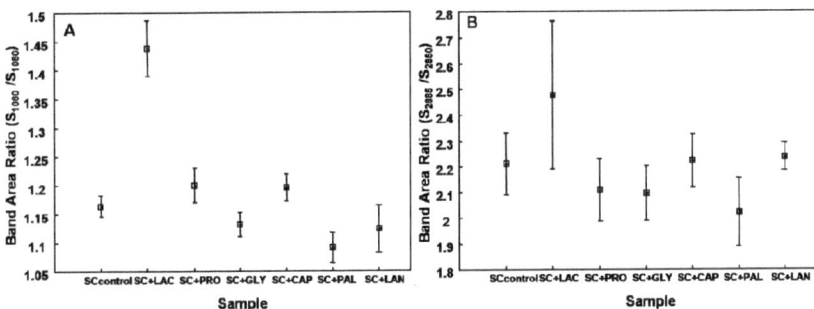

Figure 7: Conformational order of sc lipids as function of applied moisturizers after drying process (rh = 4%). The conformational changes were estimated using trans/gauche ratios : a) ν c-c vibrations (s_{1060} /s_{1080}) ; b) ν c-h vibrations (s_{2885} /s_{2850})

For samples treated with caprilic acid and LAN, 4 hours after incubation at RH 4%, the lipids conformational order appears to be equal to the control lipids conformation content. PAL, GLY and PRO provide a small trans/gauche ratio of the lipids hydrocarbon chain. Thus, those molecules are thought to disturb the lipid organization and to lead to a less ordered lipid matrix (Fig. 7). LAC offers a high trans/gauche ratio. Then, we can say that LAC increases the skin barrier function by favoring lipids conformational changes giving to the SC a more compact structure. This small molecule acts as an organization enhancer for lipids matrix.

It is important to notice that LAC increased bound water content in the SC. The interaction of bound water with the SC components may improve the lipid lamellar organization. In fact, Nakagawa et al. have shown that, lactate in the SC may play roles in maintaining the physical properties of the SC in healthy subjects more than other NMF [40].

Penetration kinetics of chemicals

Even though the main action of moisturizers is limited to SC, different components are thought to penetrate to deeper layers. ATR-FTIR investigations allowed studying the kinetics of penetration of various molecules across the SC tissues.

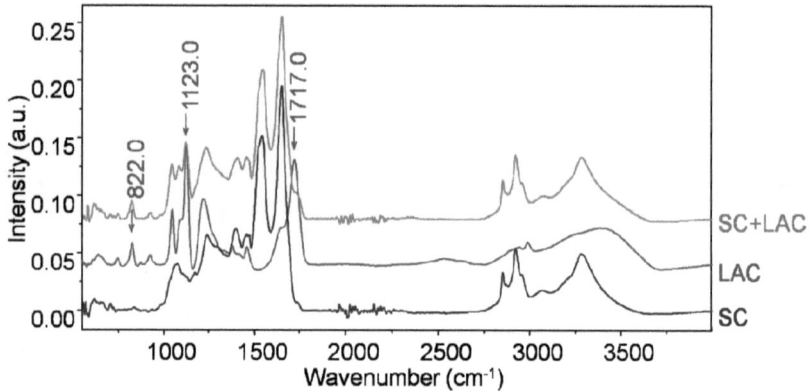

Figure 8: Comparison of the infrared spectrum of sc stuck on the atr plate, the spectrum of pure lac and the spectrum of sc 60 minutes after topical application of lac.

Fig. 8 shows an example of spectra that are obtained in a penetration study of LAC through the SC by IR spectroscopy (ATR). 8-10 minutes after it was topically applied on SC, specific peaks of the LAC were detected at 822, 1123 and 1717 cm^{-1}.

Molecule	SC Penetration	Penetration Time (min)	logP
LAC	Yes	8 -10	-0.62
PRO	No	/	-1.093
GLY	No	/	-1.76
CAP	Yes	26 - 30	-
PAL	No	/	10.82
LAN	No	/	-

Table 2: Penetration of different molecules through the *stratum corneum* investigated by ATR-FTIR

As illustrated in table 2, only 2 agents entirely crossed the SC 2 hours after their application. Even though GLY and PRO are assumed to penetrate the SC in the literature, they are usually tested in aqueous solutions which could act as a penetrating enhancer [3, 41]. Indeed, in our study, this was verified by monitoring the penetration profile of different solution of glycerol in water (0:100, 10:90, 20:80, 30:70, 40:60, 50:50, 60:40, 70:30, 80:20, 90:10, 100:0) (%). The results showed that only the 3 first mixtures (up to glycerol: water, 20:80) penetrated entirely the SC. Moreover, it has been shown that 2-butoxyethanol penetrates the skin markedly faster when dissolved in water as compared to ethanol [42] Thus, as the present used compound were pure chemicals, their diffusions appeared to be lower.

The diffusing molecules interact with SC components. This interaction could results in modification of the supramolecular organization of lipids and proteins. According to their polarity and viscosity, the non-penetrating molecules could act as occlusive.

Discussion

Application of moisturizers on the skin affects the SC structure and function (Table 3). The applied molecules modify the water content of the SC and /or change the shape of the water loss.

| samples | Water loss | 4% | | βsheet/ α-helix | | Trans/gauche | |
		Partially Bound water	Unbound water	1672/1656	1249/1280	1060/1080	2885/2850
LAC	S+	S+	NS	S+	S+	S+	S+
PRO	S-	NS	NS	S+	NS	NS	NS
GLY	S-	NS	NS	S+	NS	NS	NS
CAP	NS	NS	NS	S+	NS	NS	NS
PAL	S+	NS	NS	S+	NS	S-	S-
LAN	S+	S+	S+	S+	NS	NS	NS

Table 3: Effect of different molecules on SC investigated by Raman spectroscopy and t-test statistical analysis, S+: increasing significant effect, S-: decreasing significant effect, NS: non-significant effect.

At highly humid environment (RH=90%), some moisturizers (LAN, PAL) seem to dehydrate the SC others (GLY, PRO, LAC) increase the SC hydration level compared to the control. Previously, it has been shown that application of undiluted molecules may results in dehydration of the skin, based on osmotic water extraction from the SC [43]. Since our molecules are pure emollients, the same phenomenon could occur for some compound like LAN and PAL.

The evaluation of water retention capacity of applied molecules was based on changes in residual water content of the SC in dry environment. There are different mechanisms by which moisturizers improve barrier function. It was thought for a long time that moisturizers increase SC hydration by only two different mechanisms: by occlusion of the skin surface and by introduction of humectants, which are able to bind water in the SC. Recent studies have demonstrated that some externally applied compounds do intercalate into the skin. Those compounds can change the conformation and/or lamellar organization of the intercellular lipid matrix in the SC layers make them less permeable to water [11, 44]. Occlusion of the surface with hydrophobic substances results in reduced water loss from the outside of the skin. LAN has that property and provides to the SC a good retention capacity. It is worthy to notice that PAL appears to increase slightly the water content (Table 3). This molecule may have weak occlusive properties.

By definition, moisturizers that act as humectants will associate water into the SC but not prevent the hydrated SC from losing its increased water content. Some moisturizers (GLY, PRO, and LAC) increase the SC water caption at high level of RH (90%). However the residual water content of sample treated with GLY and PRO was not significantly different from the control. Thus, these molecules were proved to act as humectants. Meanwhile, in addition to humectants properties, LAC ensured high residual water content at RH=4% and consequently good retention capacity of water. This molecule has double action.

Besides differences in water-binding capacity of these substances, their absorption into the skin is important for their action mode determination. LAC and caprilic acid was proved to penetrate the SC within 10 an 30 minutes respectively while the others did not cross the SC tissue in the 2 hours of experiment. This difference could explain the greater effect of LAC on the lipid and protein structures. LAC is considered as humectants. However, its effectiveness to improve the SC barrier was observed. The LAC penetrates the skin and diffuses in the whole depth of SC. Thus, this component may accomplish its hydrating effect by more complex mechanisms than just by water binding [45]. The LAC molecules may interact with the SC lipid structures or proteins, improving their water-binding properties. The interaction between LAC and SC lipids is suggested to enhance skin barrier properties, and in this way to improve the water-holding capacity of the SC, which results in more effective moisturization of the skin. This is in accordance with previous findings suggesting that α-hydroxy acids such as LAC might influence the lateral packing of the lipid bilayer. They thereby help maintaining good barrier function [45-47]. In addition to that, LAC increased the bound water content. The high level of partially bound water increases the global organization of the inter-cellular lipids [16]. Thus LAC is thought to cumulate two effects: lipid modulating and humectants effects.

The present used approach should not only increase our understanding of the molecular mechanisms implicated in skin-product interactions and their clinical consequences but also allow a more solid link between skin physiopathology and therapeutic treatments thanks to an improved knowledge on the molecules structure-activity relationship. This work demonstrated that using appropriate experimental methods and data processing methods, non-invasive vibrational spectroscopy could extensively be used in dermatological sciences for characterization of skin supra-molecular organization and may help to make an effective therapeutic concept dealing with the various skin dysfunctions.

Acknowledgments:

The authors thank project: ANR-12-JSV5-0003 CARE for the financial support

References

[1] M. Denda, J. Sato, Y. Masuda, T. Tsuchiya, J. Koyama, M. Kuramoto, P.M. Elias, K.R. Feingold, Exposure to a Dry Environment Enhances Epidermal Permeability Barrier Function, 111 (1998) 858-863.

[2] J. Sato, M. Denda, S. Chang, P.M. Elias, K.R. Feingold, Abrupt Decreases in Environmental Humidity Induce Abnormalities in Permeability Barrier Homeostasis, 119 (2002) 900-904.

[3] C.R. Behl, G.L. Flynn, T. Kurihara, N. Harper, W. Smith, W.I. Higuchi, N.F.H. Ho, C.L. Pierson, Hydration and Percutaneous Absorption: 1. Influence of Hydration on Alkanol Permeation Through Hairless Mouse Skin, J Investig Dermatol, 75 (1980) 346-352.

[4] R.R. Warner, K.J. Stone, Y.L. Boissy, Hydration disrupts human stratum corneum ultrastructure, J Invest Dermatol, 120 (2003) 275-284.

[5] S. Verdier-Sevrain, F. Bonte, Skin hydration: a review on its molecular mechanisms, J Cosmet Dermatol, 6 (2007) 75-82.

[6] M.B. Jennings, L. Logan, D.M. Alfieri, C.F. Ross, S. Goodwin, C. Lesczczynski, A comparative study of lactic acid 10% and ammonium lactate 12% lotion in the treatment of foot xerosis, J Am Podiatr Med Assoc, 92 (2002) 143-148.

[7] M. Loden, A.C. Andersson, C. Andersson, T. Frodin, H. Oman, M. Lindberg, Instrumental and dermatologist evaluation of the effect of glycerine and urea on dry skin in atopic dermatitis, Skin Res Technol, 7 (2001) 209-213.

[8] C.W. Blichmann, J. Serup, A. Winther, Effects of single application of a moisturizer: evaporation of emulsion water, skin surface temperature, electrical conductance, electrical capacitance, and skin surface (emulsion) lipids, Acta Derm Venereol, 69 (1989) 327-330.

[9] M. Loden, The increase in skin hydration after application of emollients with different amounts of lipids, Acta Derm Venereol, 72 (1992) 327-330.

[10] J.M. Wiechers, T. Barlow, Skin moisturisation and elasticity originate from at least two different mechanisms, Int J Cosmet Sci, 21 (1999) 425-435.

[11] J. Caussin, G.S. Gooris, J.A. Bouwstra, FTIR studies show lipophilic moisturizers to interact with stratum corneum lipids, rendering the more densely packed, Biochimica et Biophysica Acta (BBA) - Biomembranes, 1778 (2008) 1517-1524.

[12] N. Garmy, N. Taieb, N. yahi, Interaction of cholesterol with sphingosine: physicochemical characterization and impact on intestinal absorption, J Lipid Res, 46 (2005) 36-45.

[13] J. Vilaplana, J. Coll, C. Trullas, A. Azon, C. Pelejero, Clinical and non-invasive evaluation of 12% ammonium lactate emulsion for the treatment of dry skin in atopic and non-atopic subjects, Acta Derm Venereol, 72 (1992) 28-33.

[14] M. Lodén, The skin barrier and use of moisturizers in atopic dermatitis, Clinics in Dermatology, 21 (2003) 145-157.

[15] T. Vergou, S. Schanzer, H. Richter, R. Pels, G. Thiede, A. Patzelt, M.C. Meinke, W. Sterry, J.W. Fluhr, J. Lademann, Comparison between TEWL and laser scanning microscopy measurements for the in vivo characterization of the human epidermal barrier, Journal of Biophotonics, 5 (2012) 152-158.

[16] R. Vyumvuhore, A. Tfayli, H. Duplan, A. Delalleau, M. Manfait, A. Baillet-Guffroy, Effects of atmospheric relative humidity on Stratum Corneum structure at the molecular level: ex vivo Raman spectroscopy analysis, Analyst, 138 (2013) 4103-4111.

[17] P.J. Caspers, G.W. Lucassen, G.J. Puppels, Combined in vivo confocal Raman spectroscopy and confocal microscopy of human skin, Biophys J, 85 (2003) 572-580.

[18] L. Chrit, P. Bastien, B. Biatry, J.T. Simonnet, A. Potter, A.M. Minondo, F. Flament, R. Bazin, G.D. Sockalingum, F. Leroy, M. Manfait, C. Hadjur, In vitro and in vivo confocal Raman study of human skin hydration: assessment of a new moisturizing agent, pMPC, Biopolymers, 85 (2007) 359-369.

[19] M. Egawa, T. Hirao, M. Takahashi, In vivo estimation of stratum corneum thickness from water concentration profiles obtained with Raman spectroscopy, Acta Derm Venereol, 87 (2007) 4-8.

[20] M. Egawa, H. Tagami, Comparison of the depth profiles of water and water-binding substances in the stratum corneum determined in vivo by Raman spectroscopy between the cheek and volar forearm skin: effects of age, seasonal changes and artificial forced hydration, Br J Dermatol, 158 (2008) 251-260.

[21] M. Gniadecka, O.F. Nielsen, S. Wessel, M. Heidenheim, D.H. Christensen, H.C. Wulf, Water and protein structure in photoaged and chronically aged skin, J Invest Dermatol, 111 (1998) 1129-1133.

[22] G. Zeng, J.J. Shou, K.K. Li, Y.H. Zhang, In-situ confocal Raman observation of structural changes of insulin crystals in sequential dehydration process, Biochim Biophys Acta, 1814 (2011) 1631-1640.

[23] T. Lefèvre, M.-E. Rousseau, M. Pézolet, Protein Secondary Structure and Orientation in Silk as Revealed by Raman Spectromicroscopy, Biophysical journal, 92 (2007) 2885-2895.

[24] P.J. Caspers, G.W. Lucassen, R. Wolthuis, H.A. Bruining, G.J. Puppels, In vitro and in vivo Raman spectroscopy of human skin, Biospectroscopy, 4 (1998) S31-39.

[25] A. Tfayli, E. Guillard, M. Manfait, A. Baillet-Guffroy, Thermal dependence of Raman descriptors of ceramides. Part I: effect of double bonds in hydrocarbon chains, Anal Bioanal Chem, 397 (2010) 1281-1296.

[26] E. Guillard, A. Tfayli, M. Manfait, A. Baillet-Guffroy, Thermal dependence of Raman descriptors of ceramides. Part II: effect of chains lengths and head group structures, Anal Bioanal Chem, 399 1201-1213.

[27] S. Wartewig, R. Neubert, W. Rettig, K. Hesse, Structure of stratum corneum lipids characterized by FT-Raman spectroscopy and DSC. IV. Mixtures of ceramides and oleic acid, Chemistry and Physics of Lipids, 91 (1998) 145-152.

[28] A. Tfayli, E. Guillard, M. Manfait, A. Baillet-Guffroy, Raman spectroscopy: feasibility of in vivo survey of stratum corneum lipids, effect of natural aging, Eur J Dermatol, 22 (2012) 36-41.

[29] M. Nicollier, P. Agache, J.L. Kienzler, R. Laurent, R. Gibey, N. Cardot, J.C. Henry, Action of trypsin on human plantar stratum corneum. An ultrastructural study, Arch Dermatol Res, 268 (1980) 53-64.

[30] L.C. Morejohn, J.N. Pratley, Differential effects of trypsin on the epidermis of Rana catesbeiana. Observations on differentiating junctions and cytoskeletons, Cell Tissue Res, 198 (1979) 349-362.

[31] V.S. Thakoersing, M.O. Danso, A. Mulder, G. Gooris, A. El Ghalbzouri, J.A. Bouwstra, Nature versus nurture: does human skin maintain its stratum corneum lipid properties in vitro?, Experimental Dermatology, 21 (2012) 865-870.

[32] M. Förster, M.A. Bolzinger, M.R. Rovere, O. Damour, G. Montagnac, S. Briançon, Confocal Raman microspectroscopy for evaluating the stratum corneum removal by 3 standard methods, Skin Pharmacol Physiol, 24 (2011) 103-112.

[33] K. Biniek, K. Levi, R.H. Dauskardt, Solar UV radiation reduces the barrier function of human skin, Proceedings of the National Academy of Sciences, 109 (2012) 17111-17116.

[34] S. Tfaili, C. Gobinet, G. Josse, J.-F. Angiboust, M. Manfait, O. Piot, Confocal Raman microspectroscopy for skin characterization: a comparative study between human skin and pig skin, Analyst, 137 (2012) 3673-3682.

[35] A. Savitzky, M.J.E. Golay, Smoothing and differentiation of data by simplified least squares procedures, Anal. Chem., 36 (1964) 1627-1639.

[36] J. Zhao, H. Lui, D.I. McLean, H. Zeng, Automated autofluorescence background subtraction algorithm for biomedical Raman spectroscopy, Appl Spectrosc, 61 (2007) 1225-1232.

[37] C.L. Erickson, M.J. Lysaght, J.B. Callis, Relationship between digital filtering and multivariate regression in quantitative analysis, Analytical Chemistry, 64 (1992) 1155A-1163A.

[38] N. Nakagawa, S. Naito, M. Yakumaru, S. Sakai, Hydrating effect of potassium lactate is caused by increasing the interaction between water molecules and the serine residue of the stratum corneum protein, Experimental Dermatology, 20 (2011) 826-831.

[39] E. Sprecher, G. Yosipovitch, R. Bergman, D. Ciubutaro, M. Indelman, E. Pfendner, L.C. Goh, C.J. Miller, J. Uitto, G. Richard, Epidermolytic Hyperkeratosis and Epidermolysis

Bullosa Simplex Caused by Frameshift Mutations Altering the V2 Tail Domains of Keratin 1 and Keratin 5, J Investig Dermatol, 120 (2003) 623-626.

[40] N. Nakagawa, S. Sakai, M. Matsumoto, K. Yamada, M. Nagano, T. Yuki, Y. Sumida, H. Uchiwa, Relationship between NMF (lactate and potassium) content and the physical properties of the stratum corneum in healthy subjects, J Invest Dermatol, 122 (2004) 755-763.

[41] T. Marjukka Suhonen, J.A. Bouwstra, A. Urtti, Chemical enhancement of percutaneous absorption in relation to stratum corneum structural alterations, J Control Release, 59 (1999) 149-161.

[42] H.C. Broding, A. van der Pol, J. de Sterke, C. Monse, M. Fartasch, T. Bruning, In vivo monitoring of epidermal absorption of hazardous substances by confocal Raman micro-spectroscopy, J Dtsch Dermatol Ges, 9 (2011) 618-627.

[43] J.W. Fluhr, R. Darlenski, C. Surber, Glycerol and the skin: holistic approach to its origin and functions, Br J Dermatol, 159 (2008) 23-34.

[44] J. Caussin, E. Rozema, G.S. Gooris, J.W. Wiechers, S. Pavel, J.A. Bouwstra, Hydrophilic and lipophilic moisturizers have similar penetration profiles but different effects on SC water distribution in vivo, Exp Dermatol, 18 (2009) 954-961.

[45] T. Sugawara, K. Kikuchi, H. Tagami, S. Aiba, S. Sakai, Decreased lactate and potassium levels in natural moisturizing factor from the stratum corneum of mild atopic dermatitis patients are involved with the reduced hydration state, Journal of dermatological science, 66 (2011) 154-159.

[46] L.D. Rhein, F.A. Simion, C. Froebe, J. Mattai, R.H. Cagan, Development of a stratum corneum lipid model to study the cutaneous moisture barrier properties, Colloids and Surfaces, 48 (1990) 1-11.

[47] S.G. Alderson, M.D. Barratt, J.G. Black, Effect of 2-hydroxyacids on guinea-pig footpad stratum corneum: mechanical properties and binding studies, Int J Cosmet Sci, 6 (1984) 91-100.

CHAP VIII.   CARACTERISATION *IN VIVO* DE LA PEAU PAR SPECTROSCOPIE RAMAN

ARTICLE 5 : QR CODE DE LA PEAU: EMPREINTES DIGITALES *IN VIVO* PAR
SPECTROSCOPIE RAMAN

Contexte: La santé et la fonction physiologique de la peau dépendent de nombreux paramètres. Les dermatologues et les scientifiques en cosmétique sont très intéressés par la caractérisation globale de l'état de la peau *in vivo*. En raison de la complexité de sa structure et ses fonctions, un paramètre seul n'est pas suffisant pour décrire entièrement l'état physiologique de la peau. Dans cette optique, différentes méthodes non-invasives pour le suivi des fonctions de la peau ont été mis en œuvre, offrant les avantages d'être précises et inoffensives. Ainsi, l'analyse non-destructive des propriétés physico-chimiques de l'épiderme *in vivo* est devenue accessible et a donné la possibilité d'introduire une approche multi- paramétrique pour l'évaluation de la santé du SC. Le but de cette étude a été de démontrer le principe et la preuve de concept d'une méthode quantitative *in vivo* pour différentes caractéristiques de la peau en utilisant uniquement la micro-spectroscopie Raman confocale et le traitement des données PLS . Ce travail a abouti à la proposition d'une analyse prédictive indirecte fondée sur un spectre Raman en tant que nouvel outil de recherche multi-paramétrique *in vivo*.

Méthodes: La microspectroscopie Raman confocale *in vivo* couplée à des calculs chimiométriques a été utilisée comme méthode unique pour l'étude des différentes caractéristiques de la peau. Les analyses semi- quantitatives des caractéristiques de la peau ont été réalisées sur 11 femmes volontaires présentant une peau normale. Le contenu de classes de lipides : cholestérol (CHOL), céramides (CER), acides gras (AG), le niveau d'hydratation, la perte insensible en eau (PIE) et le pH ont été quantifiés et directement corrélés au signal Raman des 5 premiers micromètres superficiels de la peau . En plus, les signatures Raman ont été utilisées pour caractériser le profil des différents types d'eau et la structure de la kératine.

Résultats: La méthodologie utilisée dans cette étude combinant les mesures Raman *in vivo* et la régression PLS a révélé une corrélation entre le signal Raman et les autres paramètres de la peau. Les spectres Raman sont caractéristiques de la structure chimique de l'échantillon étudié. Pour chaque paramètre de la peau, les modèles PLS obtenus présentent des coefficients de corrélation supérieur à $R2 = 0,99$ avec une bonne capacité de prédiction des modèles. Ainsi, la prédiction de ces caractéristiques de la peau peut être obtenue en utilisant une matrice constituée par l'ensemble des coefficients PLS calculés pour chaque observation Y, à savoir: le pH, la PIE, le taux d'hydratation, la quantité de CER, AG et CHOL et contient également des informations sur le spectre moyen et l'écart type respectivement (Xm) et Xstd ainsi que la Ym et Ystd de chaque observation à partir du modèle. Cette structure est appelée «QR code de la peau ».

De plus, nous avons constaté que les caractéristiques étudiées sont liées entre elles au niveau cutané. Nos résultats suggèrent que la diminution du niveau de pH augmente la fonction de la barrière épidermique (abaissement de PIE). Cependant, nous avons observé une évolution non linéaire d'acides gras, de PIE et d'hydratation. L'hydratation maximale et la PIE minimale sont observées lorsque la quantité d'acides gras est d'environ 40 % du total des lipides. Par conséquent, comme il est possible grâce au «QR code de la peau» d'obtenir les valeurs des différentes caractéristiques de la peau, y compris sur les quantités de lipides du SC, nous pourrions caractériser robustement l'état de la peau en combinant tous ces paramètres.

Des informations supplémentaires ont également été obtenues en profondeur grâce au système confocal Raman. Globalement, la teneur en eau augmente avec la profondeur. La fraction des zones spectrales entre 3245-3420 cm^{-1} relatives à la teneur en eau partiellement liée diminue avec la profondeur tandis fraction de la zone spectrale entre 3420-3620 cm^{-1} (eau non liée) augmente. Concernant les protéines, l'augmentation de la profondeur était accompagnée d'un processus de déroulement de la protéine (décalage de $v_{asym}CH3$ vers des nombres d'ondes plus élevés).

Conclusion: L'approche actuelle offre un grand avantage en termes d'économie de temps d'analyse et de matériel, de prévision et de compression de données multi-paramétriques. L'avantage de la méthode proposée réside dans la rapidité et l'innocuité de la mesure par spectroscopie Raman. En effet, la quantification indirecte de la teneur en lipides peut être réalisée en moins de 1 min au lieu de plusieurs heures et sans consommation de solvant qu'exige la méthode HPLC. On peut aussi obtenir des informations importantes sur le profil de composantes du SC. Avec une telle méthode rapide et simple, la corrélation entre les différents paramètres peut évaluer la sévérité de différents phénomènes physiopathologiques. Cet outil pourrait donc être utilisé pour le diagnostic en dermatologie.

ARTICLE 5

Raman spectroscopy: *in vivo* Quick response "QR" code of skin physiological status

Raoul Vyumvuhore[a], Ali Tfayli[a], Olivier Piot[b], Maud Le Guillou[c], Nathalie Guichard[c], Michel Manfait[b], Arlette Baillet-Guffroy[a].

[a]*Group of Analytical Chemistry of Paris-Sud (GCAPS), Faculty of Pharmacy, Univ. Paris-Sud, Chatenay-Malabry, France*

[b]*MéDIAN-"Biophotonics and Technologies for Health", CNRS FRE3481 MEDyC, Faculty of Pharmacy, Univ. Reims Champagne Ardennes, Reims, France*

[c]*Research & Development Department, SILAB, BP 213, Brive Cedex, France.*

Abstract

Dermatologists need to combine different clinically relevant characteristics for a better understanding of the skin health. These characteristics are usually measured by different techniques and some of them are highly time-consuming.

Therefore, a predicting model based on Raman spectroscopy and Partial Least Square regression was developed as a rapid multi-parametric method. The Raman spectra collected from the five uppermost micrometers of 11 healthy volunteers were fitted to different skin characteristics measured by independent appropriate methods (Trans-Epidermal Water Loss, hydration, pH, relative amount of ceramides, fatty acids and cholesterol). For each parameter, the obtained PLS model presented correlation coefficients higher than $R2=0.9$. This model enables to obtain all the aforementioned parameters directly from the unique Raman signature. In addition to that, in depth Raman analyses down to 20 μm showed different balance between partially bound water and unbound water with depth. In parallel, the increase of depth was followed by an unfolding process of the proteins.

The combinations of all these information lead to a multi-parametric investigation which better characterize the skin status. Raman signal can thus be used as a Quick Response code "QR code". This could help dermatologic diagnosis of physiological variations and present a possible extension to pathological characterization.

Keywords: in vivo skin analysis, Raman micro-probe, Partial Least Square, QR code, multi-parametric analysis.

Introduction

Skin covers the body and accomplishes multiple defensive functions. Its physiological health and function depends on many parameters. The epidermis represents a barrier against the loss of water and other components from the organism (1, 2). Furthermore, this epidermal barrier mainly due to the stratum corneum (SC) is the limiting unit for the penetration of exogenous substances through the skin (3-5). This well controlled function is fully assumed in intact skin, but sometimes the SC barrier status can be modified by different external factors such as climate (6), physical stressors, and internal physiological characteristics like hormonal secretion (7), nutrition as well as the composition of a lipid matrix i.e. ceramides, free fatty acids, cholesterol and its derivatives in the intercellular spaces (8-10). The modification of one parameter perturbs the skin homeostasis (11-13). Thus, the effect of each internal characteristic and its relation with the surrounding should be understood.

In that optic, dermatologists and cosmetic scientists are highly interested in global characterization of the skin status. Due to the complexity of its structure and functions, a single parameter is not sufficient to describe entirely the **skin physiological status** (14). Therefore, different methods for monitoring skin functions have been introduced.

Among the *in vivo* skin characterization parameters, transepidermal water loss, skin surface acidity (pH), SC hydration, Natural Moisturizing Factors (NMF) content and lipids classes' content are the most frequently analyzed.

The measurements of the transepidermal water loss (TEWL) gives an idea on the SC barrier function, thus providing information on permeability barrier status under normal, experimentally perturbed, or diseased conditions (15-17). Moreovever, it has been established that pH value of the skin surface plays an important role for the epidermal barrier (11, 18, 19).

SC hydration is not only important for maintaining skin functional properties but also has great impact on the skin's aesthetic properties. Different *in vivo* methods for the assessment of SC hydration have been described, namely, electrical, microwave, thermal, microscopic, magnetic resonance and spectroscopic, including Raman spectroscopy (20-24).

However, most commonly applied methods are based on measuring the electrical conductance, capacitance, or impedance as an indirect indicator for SC water content. Meanwhile, those conventional classical *in vivo* noninvasive methods do not provide direct information about the depth profiles of water content, and SC components structure.

Confocal Raman microspectroscopy is the first commercially technique that provides an *in vivo* non-invasive method to determine depth profiles of water concentration in the skin (20, 25-30). In-depth measurements allow biopharmaceutical studies (31-33). This technique can be also used to study skin physiology and pathological conditions including cancers (34-38).

Thus, Raman spectroscopy seems to be highly sensitive technique to tiny molecular changes. Therefore, one can assume that variations in a skin parameter would induce changes in the vibrational state. Raman signature associated to the adequate data processing appears to be a good candidate for multi-parametric analysis of the skin.

Partial Least Square (PLS) processing coupled to Raman spectroscopy has been presented as a novel approach making possible to obtain qualitative information about the distribution of the compounds (39, 40).

The aim of this study was to demonstrate the proof-of-principle of a novel and **single quantitative method** *in vivo* assay for different skin characteristics by using confocal Raman micro-spectroscopy and PLS data processing.

In that purpose, semi-quantitative analyses of skin characteristics were realized on volunteers with normal non-diseased skin. Lipid classes' contents: Cholesterol, ceramides, fatty acids; hydration level; TEWL and pH were monitored and directly correlated to the Raman signal of the five outermost micrometers of the skin. The use of the different fitting coefficients in the validation set enabled to obtain all the different parameters directly from the unique Raman signature. Moreover, in addition to information obtained from the PLS models, in depth Raman analyses enabled to evaluate the variation in water and protein structure in deeper layers. The combinations of all these information lead to a multi-parametric investigation which better characterizes the skin status. Raman signal can thus be used as a skin Quick Response code "skin QR code", **because the information in a Raman spectrum can be decoded quickly** for predicting different skin characteristics.

Materials and methods

The *in vivo* experiments were performed on 11 healthy volunteers and measurements were undertaken at 4 sampling locations: the volar and outer forearm as well as inside and outside calf. After a rest period of 20 min in an environmental controlled room, the different physiological parameters were measured on the same side of the skin.

In vivo **Raman microprobe analysis**

In vivo Raman investigations were performed using an *in vivo* confocal Raman optical microprobe (Horiba Jobin Yvon, Villeneuve d'Asc, France).

The probe is coupled to 5 μm diameter fiber optic probe (InPhotonics, USA) with a coaxial two-fiber probe, one for excitation and the other for collection. A band pass filter removes the unwanted signals transmitting pure laser light. A double pumped Nd:YLF laser source at 660 nm (Toptica Photonics, Munich, Germany), with 12 mW power, on the sample, was used. The 660 nm excitation wavelength was chosen because it gives a weak background in the fingerprint region (41) and a high Raman Stokes signal in the high wavenumbers region. A long focal microscope objective MPlan FLN 100X / NA 0.75 WD 4.7 (Leica, Mannheim, Germany) (3 μm in-depth resolution) was coupled to a piezoelectric system allowing "z" movement of 100 microns with adjustable increment. After the lens, an independent device with aperture of 1 or 2 mm, for adjusting the distance between the skin and the lens was added to the system so as to be positioned at the optimum working distance. An ultrafast autofocus system allows focusing the laser on the surface of the skin in near real time. A display system for viewing the surface of the sample is also incorporated to the probe.

Collected scattered light was analyzed with Micro-HR Raman spectrometer with a compact rugged geometry (178x267x140 mm) and no internal moving parts. The raw signal was filtered through an edge filter and then dispersed using a 600 lines/mm holographic grating and a 100 microns slit. The Raman Stokes signal was recorded with a high sensitivity CCD detector cooled to -70 ° C by Peltier effect. The dispersion of the signal provides a total coverage from 400 to 3700 cm^{-1} and spectral resolution of 1 cm^{-1}. The used system is equipped by a confocal micro-probe attached to a movable arm allowing to position easily the probe on skin surface.

To monitor the intra-individual variability of the skin physiological status each location of each volunteer was analyzed on 7 different points. For each analyzed point, Z-profiles are made by moving the lens along the optical axis from 0 to 20 μm depths with a 2 μm step. This gives depth information of the SC in vivo. All spectra were subjected to the same automated pre-processing protocol. All spectra were smoothed using a Savitsky-Golay algorithm on 11 points (42). In order to avoid the interference of the fiber background in the analysis, spectra were corrected by baseline subtraction using an automatically generated polynomial function which individually fits to each spectrum of the data set (37).

The exact position of the skin surface was determined using the method described by Tfayli et al. (43). Absolute intensity variations of spectra were corrected by vector normalization.

Biometric measurements

The measured values were Trans-Epidermal water loss (TEWL) using Tewameter TM 210, pH value measured with Skin pH meter 905 and skin hydration obtained with Corneometer CM 820 all from (Courage & Khazaka electronic GmbH, Köln, Germany). To minimize the intra-individual variation, the presented TEWL, pH and skin hydration values are the mean of 3 measurements.

In vivo SC lipid extraction

The *in vivo* SC lipid extraction was performed using a previously described method (44, 45). The sampling area of the skin was wiped (without rubbing) with a filter paper moistened with ether. Then, a surface of 5x4cm was drawn. For lipid extraction, a cotton swab presoaked in a mixture of ethyl acetate / methanol (20/80 v / v) was used to skip the skin area 10 times in the same direction. The two sides in cotton were removed from cotton swab using fine forceps and placed in a glass flask with PTFE stopper. The last step was repeated 3 more times, each time a new cotton swab was used in a different rubbing direction of the rectangle. The flask containing 8 halves of four cotton swabs was surrounded with Parafilm ® M at the bottle and cap closure and then the entire bottle was wrapped with aluminum foil. The extracts were kept at -20 ° C. The contents of the vials were thawed for analysis at room temperature. 6 mL of a chloroform / methanol mixture (2:1 v / v) were added in the bottle and vortexed for 2 minutes. The solvent mixture was collected and putted it in a glass test tube before evaporation to dryness under a stream of nitrogen with gentle heating (50 ° C). The extraction is repeated with 6 mL of solvent mixture and insert after stirring the dry residue formed previously. The obtained mixture is evaporated once again in the previous described conditions. Therefore, the dry residue is dissolved in 200 µl solvent mixture of heptane/(chloroform:methanol 2:1) (180:20 v/v) before HPLC analysis.

High Performance Liquid Chromatography (HPLC) instrumentation

High performance liquid chromatography was performed using a normal phase separation (46, 47). A silica grafted polyvinyl alcohol (PVA)-Sil column (PVA-bonded column; 5 µm particle size, 150× 4.6 mm) purchased from YMC (Kyoto, Japan) and thermostated at 35°C in an Ultimate oven 3000 RS was used for separation. The solvents were degassed prior to use and the mobile phase was continuously degassed in a degasy RSLC Ultimate 3000 connected to a Dionex® Ultimate 3000 RSLC pump.

Normal-phase liquid chromatography was performed using a binary gradient solvent system of heptane/chloroform (80:20 v/v) (solvent A) and acetone (solvent B) using a flow rate of 1 mL/min and the injected volume was 5 µL. The HPLC system was coupled to Corona® CAD (ESA Biosciences, Chelmsford, MA, USA) . The analysis was performed using Chromeleon software. Solvents (Heptane (99.7% purity), chloroform (99.3% purity), and acetone (99.7% purity)) were purchased from VWR (Fontenay-sous-bois, France).

Chemometric analysis

Multivariate data analysis was performed on Simca P-11 (Umetrics, Sweden) software. Partial Least Squares (PLS) (48-50) method was applied for the identification of Raman features correlated to the different studied parameters. PLS is a multilinear calibration method that can be used for prediction of events involving multiple variables. The process calculates the maximum covariance between two matrixes X and Y. Thus, the main aim of PLS is to predict the Y-variables from the X-variables. Applied on Raman spectra (X-variables), the aim is to obtain an estimation of the multi-parametric skin characteristics (Y-variables) from the spectral data set.

The X matrix of the PLS models contained smoothed, baseline corrected and scaled spectral data using univector method. The X and Y matrix was centred and autoscaled: Unit Variance (UV) scaling which gives to all variables equal weight. The Y matrix contained the relative amount of Ceramides, Fatty Acids, Cholesterol, hydration level, TEWL, pH and different depth of the analyzed site of the skin as well.

Orthogonal Signal Correction (OSC) (51) was performed prior to Partial least square (PLS) regression.

The PLS fitting model:

For a clear illustration of how the PLS coefficient are used for prediction of the present skin characteristics, it's necessary to explain briefly how the linear PLS model is obtained (48-50): First, Principal component analysis is performed for X variables and finds new orthogonal variables that are rotated to fit with Y. They are called T-scores. They are estimated as linear combinations of the original X variables with the weights 'W'. For the model calibration we determine the appropriate number of latent variables by cross-validation technique with 7 rounds and a maximum of 200 iterations. The approach leaves out data in turn for latent variable calculation and stops when Predictive Residual Sum of Squares (PRESS) is not significantly improved. The number of latent variables was limited to 4 to avoid model over-fitting.

$$T = XW \hspace{4cm} \text{Equation 1}$$

The T-scores models X and are predictors of Y

$$Y = TC' + F \qquad \text{Equation 2}$$

C expresses the Y weights. The Y-residual matrix, F represents the deviations between the observed and modeled responses. Thus the regression model can be written in the following way:

$$Y = X\,W * C' + F = X\,B + F \qquad \text{Equation 3}$$

The PLS regression coefficients 'B' can be summarized as:

$$B = W*C' \qquad \text{Equation 4}$$

As illustrated by the Eq.4, having B and X, we can predict Y

As the X and Y data used for coefficient calculation were UV scaled; the new spectral data 'Xi' that will be used for prediction should be equally transformed. For that, the mean (Xm-vector, Ym-value) and the standard deviations (Xstd-vector, Ystd-value) of X and Y respectively from the model are needed. Thus, new 'Yi' prediction could be calculated as follows:

$$Xic=(Xi-Xm)/\,Xstd \qquad \text{Equation 5}$$

$$Yic=Xic*B \qquad \text{Equation 6}$$

Where: Xic is UV scaled Xi

Yic is UV scaled Yi

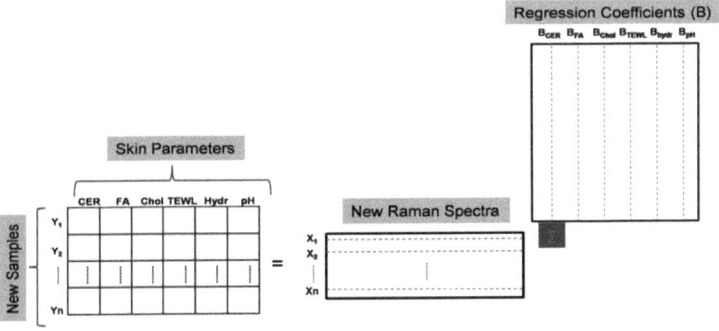

Fig. 20: Prediction of skin characteristics of (ceramides: CER, fatty acids: FA, cholesterol: Chol, hydration level, TEWL, pH) using Raman spectrum and PLS regression coefficients

Yi= Yic*Ystd+Ym   Equation 7

Thus, the proposed fitting model contains a matrix composed by all PLS coefficients (B) calculated for each Y observation: amount of ceramides, fatty acids, Cholesterol, pH, TEWL, and hydration level (Fig. 1). The fitting model's structure should also contain information about the mean spectrum (Xm) and Xstd as well as the Ym and Ystd of each observation.

Results and Discussion

Prediction of skin characteristics using SC Raman signal and PLS

The first part of this work deals with the potential of Raman spectroscopy to indirectly quantify the SC lipid classes, pH, TEWL and hydration of different volunteers. Raman spectra collected from the five outermost micrometers of the skin surface were coupled to PLS regression to monitor the different skin parameters; and try to develop a predicting model. The outermost five micrometers of the skin were chosen for PLS fitting model to avoid the influence arising from in depth variations of the skin structure.

Raman analysis can quantify indirectly ceramides, fatty acids, Cholesterol content, pH, TEWL and hydration level

In vivo Raman measurements revealed a correlation between the Raman signal and other studied parameters of the skin. Raman spectra are characteristic of the chemical structure of the studied sample. For each skin parameter (relative amount of ceramides, fatty acids, Cholesterol, pH TEWL, hydration, level), the obtained PLS model presented correlation coefficients higher than R2=0.9 reflecting good predictive ability of the models (Fig. 2).

In order to obtain an estimate of the predictive ability of the present PLS regression model, a cross validation (52, 53) was used. Some characteristics of the PLS model are illustrated in table 1:

Parameter	Compo -nents	R2X (cum)	R2X	R2VY Adj(cum)	R2VY (cum)	Q2VY (cum)	RSD (Y)	RSD (Y)WS
Ceramides	2	0.3239	0.1415	0.993	0.993	0.988	1.233	0.085

Fatty Acids	2	0.3016	0.1152	0.993	0.993	0.985	1.142	0.086
Cholesterol	3	0.2776	0.1006	0.951	0.952	0.909	1.672	0.222
pH	2	0.3563	0.1240	0.988	0.989	0.978	0.054	0.108
TEWL	4	0.3249	0.2839	0.927	0.931	0.812	0.241	0.269
Corneometry	4	0.3648	0.1897	0.908	0.913	0.755	1.767	0.303

Where :

R2X(cum) : Predictive + orthogonal variation in X that is explained by the model.

R2X : Is the amount of variation in X that is correlated to Y.

R2VY : The cumulative percent of the variation of the response explained by the model after the last component. R2 is a measure of fit, i.e. how well the model fits the data.

R2VYAdj : The cumulative percent of the variation of the response, adjusted for degrees of freedom, explained by the model after the last component.

Q2VY : The cumulative percent of the variation of the response predicted by the model, after the last component, according to cross validation. Q2 tells you how well the model predicts new data. A useful model should have a large Q2.

RSD(Y) : Residual standard deviation of Y, in original units, after the last component.

RSD(Y)WS : Residual standard deviation of Y, as scaled in the workset, after the last component.

The difference between the predicted parameters and true value were very small. Root Mean Square Error of Estimation (RMSEE) and Root Mean Square Error of Prediction (RMSEP) were found to be less than 0.5% and 0.7% respectively. The mean confidence interval for the test set was between 0.1 and 0.2, demonstrating a good precision and accuracy for those models. Thus, we showed that we can indirectly quantify the different parameters of the skin using only its Raman spectra (Fig. 2).

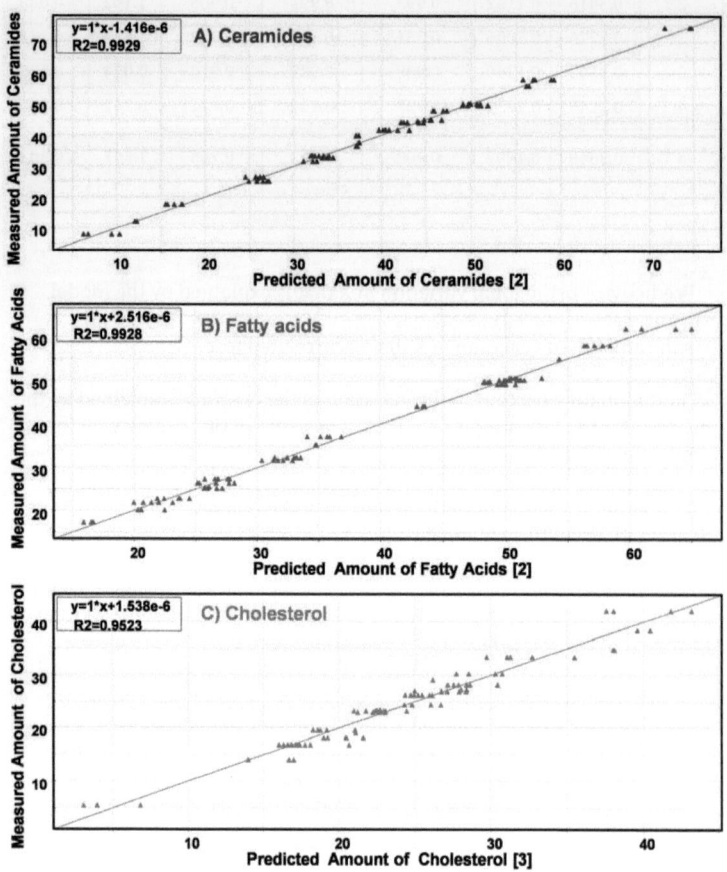

Fig. 2: PLS regression after Orthogonal Signal Correction (OSC): Amount of lipids quantified by HPLC versus predicted amount of lipids using Raman spectra and PLS data processing . The Y observation is: A) Ceramides; B) Fatty acids; C) Cholesterol. The number of latent variables is in brackets on x-axis.

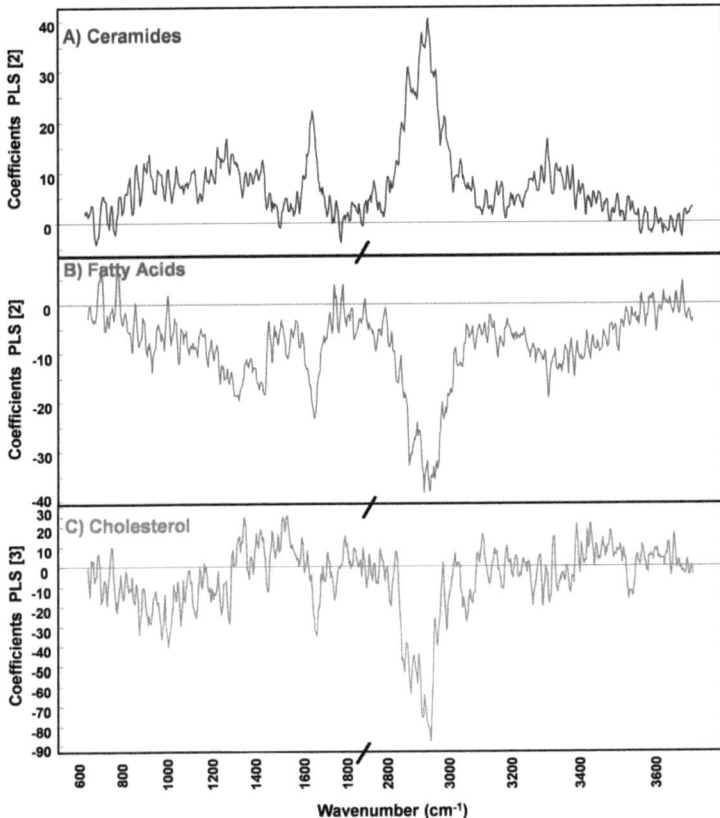

Fig. 3: Coefficient plot of PLS regression after Orthogonal Signal Correction (OSC). These coefficients correspond to the regression vectors in B matrix of Eq. 4. The Y observation is: A) Ceramides; B) Fatty acids; C) Cholesterol. The number of latent variables is in brackets on y-axis.

In addition to that, the PLS coefficient "B" (Eq. 4) showed that Raman features are sensitive to a specific analyzed parameter. The model enables the detection of those characteristic Raman modifications which were related to each studied criterion of the analyzed site. As an example, Fig. 3 illustrates the evolution of the Raman signal related to each class of SC lipids i.e. ceramides, fatty acids, cholesterol. Taking the present PLS processing, as up mentioned, it is possible to set up a **"data structure"** that contains the

PLS calibration factors and use them for prediction of newly acquired spectra (Fig. 1; Eq. 7).

Like every instrumental measurement, a calibration procedure is needed for the present method. The presently found coefficients should work on data acquired with the same Raman microprobe and by using exactly the same data pre-processing and processing. It's known that the noise and the spectral resolution depend on the performances of a system used in Raman spectroscopy measurements. The pre-processing methods used as the fluorescence removal algorithm are highly affecting the spectral features shape (54). Therefore, for a different operating environment, a new model should be developed. Thus, to facilitate clinical use, integrated real-time Raman spectroscopy systems (55) for skin evaluation and characterization, which combines real-time data acquisition and processing should be encouraged.

The present processing showed that there is a lot of information hidden in a Raman spectrum. Therefore Raman spectra could give information on hydration, TEWL, pH, as well as relative amount of ceramides, fatty acids, and cholesterol in the SC. Each of these skin characteristics is important to evaluate the skin's health.

The advantage of the proposed method lies in the rapidity,the non invasiveness, the molecular characteristics, and the information amount of the Raman spectroscopy measurement. Indeed, indirect quantification of lipid content could be done in less than 1 min instead of many hours and solvent consuming that requires HPLC method. This means that the cost of analyses is dramatically reduced. Instead of using many apparatus and specialists of all the up-mentioned analytical methods, the only Raman micro-spectrometer is needed to quantify fully all these skin parameters.

Relationship between skin characteristics

In addition to the accessibility to different characteristics of skin by combining different approaches it is of highest interest to evaluate how different skin characteristics are interrelated. Thus, the internal balance between different classes of lipids was monitored. Moreover, the measured biometric values were compared between each other in order to highlight the link involved between SC hydration level, barrier function, acidity and its lipid composition.

SC Lipid composition

The lipid composition of the epidermis governs the skin barrier, mechanics and appearance. The methods commonly used to determine lipid profiles is HPLC (46, 47, 56).

The used polar stationary phase allows separation of the lipids according to the nature or the polar head. The present used method refer to the work of Merle et al. (46). Control solutions of cholesterol, palmitic acid, cholesteryl palmitate, glyceryl trioleate and ceramides (cer2, Cer IIIB, Cer 6) were injected to determine the elution order and retention time. All the lipid classes are eluted between 2-26 min with the elution order as follow: Cholesterol ester is eluted within 2 minutes, fatty acids between 3.5 and 6 minutes, cholesterol about 5 min, it is followed by ceramides. Within the ceramides class, the elution order is essentially conditioned by the number of OH content in the polar head of ceramides Thus we observe the first elution of dihydrosphingosine bases and sphingosine bases that would be followed by phytosphingosine bases and 6-hydroxy sphingosine bases. There is a partial overlapping area between the elution characteristics of each base depending on the alkyl chain lengths and number of unsaturations. Similar elution was described by Van Smeden et al. (47) on normal phase with a column of PVA.

As example, Fig.4 represents a chromatogram obtained from one sample

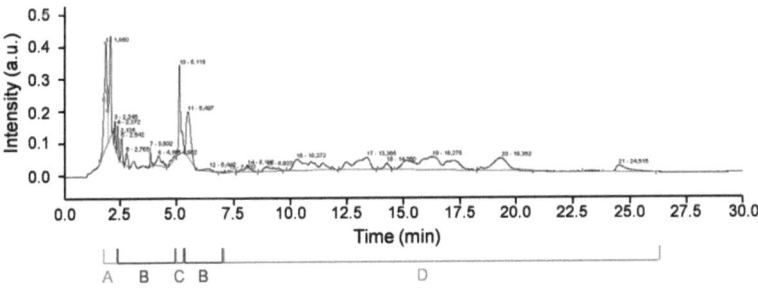

Fig. 4: Chromatogram of *in vivo* SC lipids extracts obtained using up mentioned HPLC/Corona system: A) Cholesterol Esters B) Fatty Acids, C) Cholesterol, D) Ceramides.

Relative quantification was performed by lipid classes after deducting interfering compounds present in the cotton swab areas. The results of lipid class amounts are presented as percentage of the total chromatogram area.

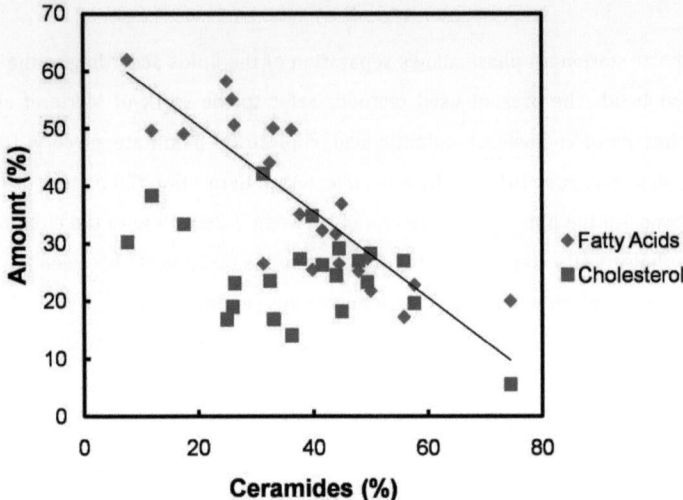

Fig. 5: Correlation between the 3 classes of SC lipids: ceramides, fatty acids, cholesterol. The quantification is expressed in percentage of the chromatogram total area.

Fig. 5 illustrates that fatty acids evolved in the opposite way from ceramides whereas the cholesterol varies slightly. Variations in skin composition and supramolecular structure may impact on biometric measurement and/or Raman spectra. It is worthy to notice that Raman features from Fig. 3 evolve in the same way for fatty acid and cholesterol while ceramides features evolve inversely. The same phenomenon is observed on Fig.5 using HPLC quantification. Once more, this observation illustrates that the PLS coefficients of the model can be used to calculate the different skin parameters based on a Raman spectrum.

Lipid content impact on hydration level and TEWL

We have seen that fatty acids are inversely related to ceramides with small variations of cholesterol amount. It must be interesting to investigate how the balance between those two types of lipids affects other skin characteristics.

Free fatty acids are thought to play a important role in the SC acidification and its role not only for barrier homeostasis but also for the dual functions of stratum corneum integrity and cohesion (19). Here no such effect on pH was demonstrated (data not shown). However, we have observed a non linear evolution of fatty acids, TEWL and hydration. Maximum hydration and min TEWL are observed when fatty acids amount is around 40 % of total lipids (Fig. 6).

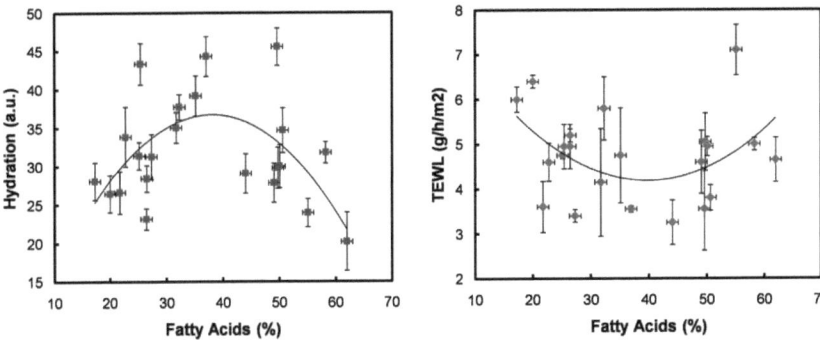

Fig. 6: Comparison between fatty acid content of the SC, hydration level and TEWL. **Each patient was analyzed on two anatomical areas.** The curves are based on data fitting using polynomial function and serve to guide the eye.

It is demonstrated that the present studied characteristics are inter-related in the skin (57). For example, the pH modifications affect the conformation of some molecules like urocanic acid UCA and thus modify the spectrum of the SC samples (58). Moreover, the influence of the lipid composition and the barrier function was highlighted based on the study of the conformational order and the compactness of the lipid matrix (35, 59, 60). Therefore, since the proposed skin QR code could provide the values of different skin characteristics including on SC lipids, we can characterize the relationship existing between all these parameters for different skin status such as; physiological status, pathological phenomena, population specificity. This multi-parametric investigation would give us a better understanding of the studied phenomena

Skin surface pH impacts on epidermal barrier function

From figure 7, we observed a trend relating TEWL and pH of the investigated skin site.

Our results suggest that a decrease of the pH level increases the epidermal barrier function (lowering TEWL). This is in accordance with previous studies. It has been established that acidic status of the skin surface plays a central role for the epidermal permeability barrier homeostasis (11), and the restoration of the disrupted barrier. The recovery of the perturbed barrier is delayed at a neutral pH, due to the disturbance in extracellular processing of the SC lipids (18). Therefore, acidification of the skin is advised in diseases with skin barrier (17).

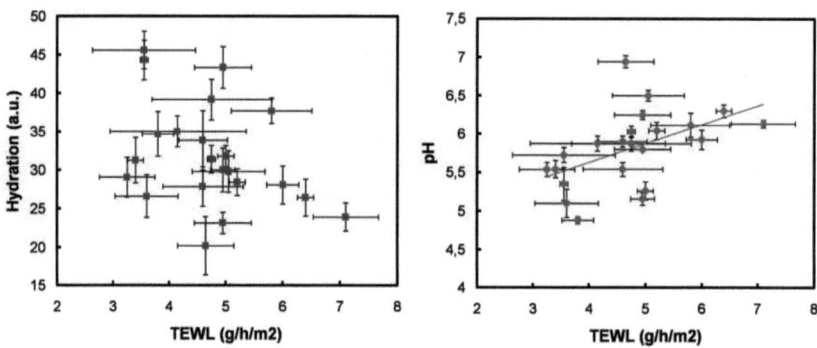

Fig. 7: Comparison TEWL, hydration level and pH values the epidermis: the curves are based on data fitting using linear function and serve to guide the eye.

Meanwhile, the hydration from corneometer measurements seemed to vary randomly (Fig. 7). Even though there is a slight trend, in contrary to previous studies (61), no correlation between SC hydration and TEWL values was observed in the present study. However, it is known that water content of the SC affects its physical properties such barrier status as well as the regulation of physiological functions. The lack of the SC water content generally induces dryness and impairs epidermal barrier function (62). Thus, it is increasingly evident that methods like corneometry are less useful in the assessment of skin dryness. Indeed, depending on the frequency used, this technique penetrates around 40µm under the skin surface which is deeper the SC layer (thickness~20µm). It has been shown recently that the only correlation between capacitance measurements and confocal Raman spectroscopy data is the water content in the lower layers of the epidermis (63). Moreover this technique is not able to detect small changes in SC hydration and is subject to interference from others substances (64). The observations appear to be supported by near infrared analysis (NIR), which indicates that impedance measurements and capacitance are relatively insensitive to small water content changes (65) Then, confocal Raman microspectroscopy appears to be an alternative and more resolute technique to record with precision the SC hydration level and water structure in the SC.

In vivo SC depth analyses

In addition to the multi-parametric characterization of the skin by combining PLS models to Raman spectra, and the interrelation between the different parameters, Raman data is directly informative on the water structure (tightly bound, partially bound and unbound water) and protein folding within the skin. In this optic, in-depth analyses were performed in order to assess evolution of water and protein structure.

Skin hydration profile

The electrical methods give an integrated value of the SC hydration, rather than the actual water distribution of the superficial epidermal layer. The efficiency of the Raman measurement to evaluate the hydration profile of Stratum corneum (SC) was determined through the correlation with high-frequency electrical conductance of tape-stripped human skin *in vivo* (66). The results obtained in the present study on water profile using Raman micro-spectroscopy revealed that global water content rises gradually from the skin surface to the lower parts of SC (data not shown). This is accordance with previously recorded data using *in vivo* confocal Raman microspectroscopy (20, 28). In addition to that, Raman microscopy can provide information on different structure of water (23, 67-70). Interestingly, as observed on **PLS coefficient plot** (Fig. 8, Eq. 4), **the fraction of spectral features between 3245-3420 cm^{-1} (zone A on Fig. 8) related to partially bound water content decreased with depth (Fig. 9A), while fraction of spectral features between 3420-3620 cm^{-1} (zone B on Fig. 8) from unbound water increased with depth (Fig. 9B).** In parallel, a recent in vitro study has revealed that over 60% of relative humidity, the SC partially bound water content decreases in favor of unbound water (23).

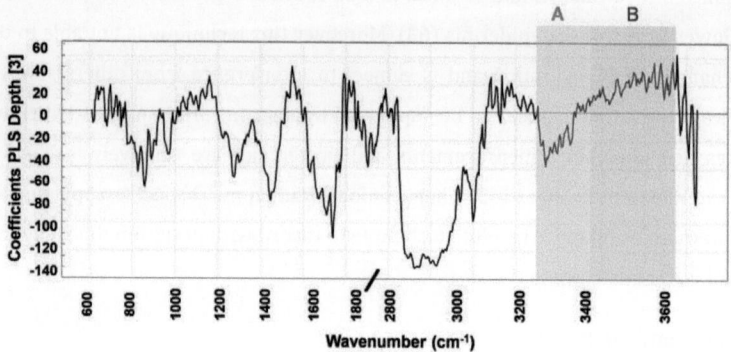

Fig. 8: Coefficient plot of PLS regression after Orthogonal Signal Correction (OSC). The Y observation is depth under skin surface down to 20 micrometers.

The observed phenomenon must be due to the equilibrium between those two types of water. The increase of unbound water weakens the forces of intermolecular bonds between water and SC components which results in a decrease of the partially bound water features around 3300 cm⁻¹. The same behavior was observed for forearm as well as for calf.

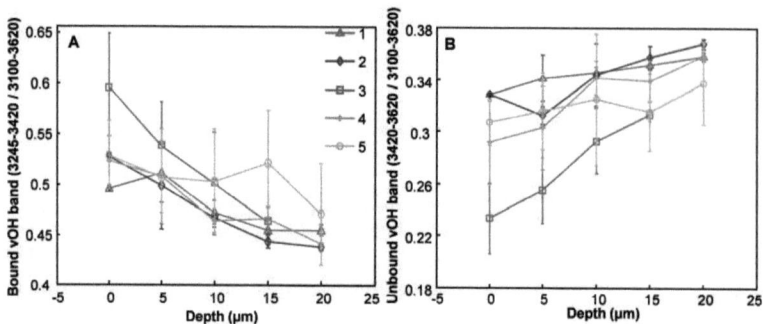

Fig. 9: Water profile of volar forearm stratum corneum obtained with *in vivo* confocal Raman microspectroscopy down to 20 μm depth of different volunteers 1, 2, 3, 4, 5. The band areas were calculated directly on the experimental spectra: A) Partially bound water fraction; B) Unbound water fraction

The unique and direct quantification of partially bound and unbound water content provided by *in vivo* confocal Raman microspectroscopy offers a whole new perspective for fundamental skin moisturization. This is an additional application of the present technique that has proved the ability to assess detailed *in vivo* concentration profiles of water and of natural moisturizing factor for the SC (20, 25-28, 71).

Protein folding profile

The modification of water content is accompanied with protein structure changes. The shift of $\nu_{asym}CH_3$ towards high wavenumbers testifies to the unfolded status of the keratin so favored due to the interaction between the side chains of its amino acids and water molecules (67, 68).

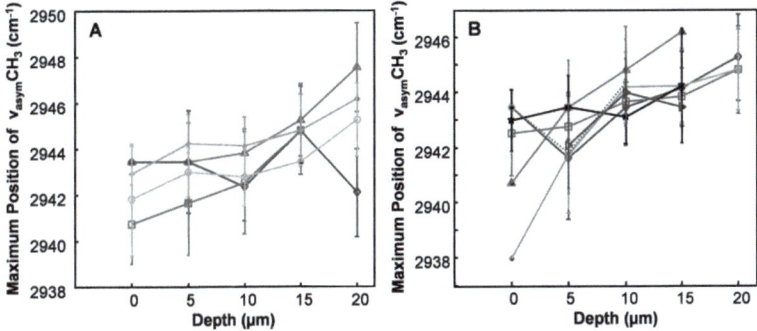

Fig. 10: Protein folding status evaluated using maximum intensity wavenumber of the peak around 2940 cm^{-1} monitored directly on the experimental spectra: *in vivo* skin Profile down to 20 μm depth: A) volar forearm; B) calf.

For all analyzed sites, the increase of depth and consequently the increase of water were followed by an unfolding process of the protein (shift of $\nu_{asym}CH_3$ towards higher wavenumbers). This is in conjunction with *ex vivo* observations (23).

Conclusion

The methodology used in the present study combines chemometrics and analytical measurements. This processing greatly advances our capabilities for research by dealing with large volumes of information that involves computer-based resources such as algorithms and multivariate statistical methods. A key task in this processing is to facilitate the representation of information hidden in data for a better understanding of the underlying biological complexities.

The present *in vivo* non invasive Raman analysis of the skin enables a very quick evaluation of a big amount of stratum corneum characteristics (quantifies ceramides, fatty acids, cholesterol, pH, TEWL, bound water profile, unbound water profile, protein folding profile). The presently developed approach offers a big advantage in terms of prediction and data compression leading in saving time and materials. This simplifies skin investigation and thereby other techniques will consider only what is not predicted by Raman measurements.

The combinations of all these obtained information lead to a multi-parametric investigation which better characterizes the skin status. **Raman signal could thus be used as a "Skin QR code"** for predicting different skin characteristics and to assess their link with different physiological and pathological phenomena. Thus, in dermo-cosmetic field, this promising and suitable method will undoubtedly offer new opportunities of skin health test evaluation.

Acknowledgments:

The authors thank project: ANR-12-JSV5-0003 CARE for the financial support

References

1.   S. E. Friberg, I. Kayali, W. Beckerman, L. D. Rhein and A. Simion, "Water permeation of reaggregated stratum corneum with model lipids," *J Invest Dermatol* 94(3), 377-380 (1990)

2.   P. M. Elias, "Epidermal barrier function: intercellular lamellar lipid structures, origin, composition and metabolism," *Journal of Controlled Release* 15(3), 199-208 (1991)

3.   V. Schreiner, G. S. Gooris, S. Pfeiffer, G. Lanzendorfer, H. Wenck, W. Diembeck, E. Proksch and J. Bouwstra, "Barrier characteristics of different human skin types investigated with X-ray diffraction, lipid analysis, and electron microscopy imaging," *J Invest Dermatol* 114(4), 654-660 (2000)

4.   G. M. Golden, D. B. Guzek, A. H. Kennedy, J. E. McKie and R. O. Potts, "Stratum corneum lipid phase transitions and water barrier properties," *Biochemistry* 26(8), 2382-2388 (1987)

5.   P. M. Elias and D. S. Friend, "The permeability barrier in mammalian epidermis," *J Cell Biol* 65(180-191 (1975)

6.   M. Denda, J. Sato, Y. Masuda, T. Tsuchiya, J. Koyama, M. Kuramoto, P. M. Elias and K. R. Feingold, "Exposure to a Dry Environment Enhances Epidermal Permeability Barrier Function," *J Invest Dermatol.* 111(5), 858-863 (1998)

7.   A. Sandilands, C. Sutherland, A. D. Irvine and W. H. McLean, "Filaggrin in the frontline: role in skin barrier function and disease," *J Cell Sci* 122(Pt 9), 1285-1294 (2009)

8.   G. Imokawa, A. Abe, K. Jin, Y. Higaki, M. Kawashima and A. Hidano, "Decreased level of ceramides in stratum corneum of atopic dermatitis: an etiologic factor in atopic dry skin?," *J Invest Dermatol* 96(4), 523-526 (1991)

9.   K. Akimoto, N. Yoshikawa, Y. Higaki, M. Kawashima and G. Imokawa, "Quantitative analysis of stratum corneum lipids in xerosis and asteatotic eczema," *J Dermatol* 20(1), 1-6 (1993)

10.   J. A. Bouwstra, F. E. Dubbelaar, G. S. Gooris and M. Ponec, "The lipid organisation in the skin barrier," *Acta Derm Venereol Suppl (Stockh)* 208(23-30 (2000)

11.   J. P. Hachem, D. Crumrine, J. Fluhr, B. E. Brown, K. R. Feingold and P. M. Elias, "pH directly regulates epidermal permeability barrier homeostasis, and stratum corneum integrity/cohesion," *J Invest Dermatol* 121(2), 345-353 (2003)

12.   A. Alonso, N. C. Meirelles, V. E. Yushmanov and M. Tabak, "Water increases the fluidity of intercellular membranes of stratum corneum: correlation with water permeability, elastic, and electrical resistance properties," *J Invest Dermatol* 106(5), 1058-1063 (1996)

13.   J. Sato, M. Denda, S. Chang, P. M. Elias and K. R. Feingold, "Abrupt Decreases in Environmental Humidity Induce Abnormalities in Permeability Barrier Homeostasis," 119(4), 900-904 (2002)

14.   Zuang, Val, rie, Rona, Claudia, Archer, Graeme, Berardesca and Enzo, *Detection of skin irritation potential of cosmetics by non-invasive measurements*, Karger, Basel, SUISSE (2000).

15.   J. d. Plessis, A. Stefaniak, F. Eloff, S. John, T. Agner, T.-C. Chou, R. Nixon, M. Steiner, A. Franken, I. Kudla and L. Holness, "International guidelines for the in vivo assessment of skin properties in non-clinical settings: Part 2. transepidermal water loss and skin hydration," *Skin Research and Technology* 19(3), 265-278 (2013)

16.   J. S. Ferguson, W. Yeshanehe, P. Matts, G. Davey, P. Mortimer and C. Fuller, "Assessment of skin barrier function in podoconiosis: measurement of stratum corneum hydration and transepidermal water loss," *British Journal of Dermatology* 168(3), 550-554 (2013)

17.   R. Darlenski, S. Sassning, N. Tsankov and J. W. Fluhr, "Non-invasive in vivo methods for investigation of the skin barrier physical properties," *Eur J Pharm Biopharm* 72(2), 295-303 (2009)

18.   K. M. Hanson, M. J. Behne, N. P. Barry, T. M. Mauro, E. Gratton and R. M. Clegg, "Two-photon fluorescence lifetime imaging of the skin stratum corneum pH gradient," *Biophys J* 83(1682-1690 (2002)

19.   J. W. Fluhr, J. Kao, M. Jain, S. K. Ahn, K. R. Feingold and P. M. Elias, "Generation of Free Fatty Acids from Phospholipids Regulates Stratum Corneum Acidification and Integrity," 117(1), 44-51 (2001)

20.   M. Egawa, T. Hirao and M. Takahashi, "In vivo estimation of stratum corneum thickness from water concentration profiles obtained with Raman spectroscopy," *Acta Derm Venereol* 87(1), 4-8 (2007)

21.   P. J. Caspers, G. W. Lucassen, E. A. Carter, H. A. Bruining and G. J. Puppels, "In vivo confocal Raman microspectroscopy of the skin: noninvasive determination of molecular concentration profiles," *J Invest Dermatol* 116(3), 434-442 (2001)

22.   P. J. Caspers, G. W. Lucassen, R. Wolthuis, H. A. Bruining and G. J. Puppels, "In vitro and in vivo Raman spectroscopy of human skin," *Biospectroscopy* 4(5 Suppl), S31-39 (1998)

23.   R. Vyumvuhore, A. Tfayli, H. Duplan, A. Delalleau, M. Manfait and A. Baillet-Guffroy, "Effects of atmospheric relative humidity on Stratum Corneum structure at the molecular level: ex vivo Raman spectroscopy analysis," *Analyst* 138(14), 4103-4111 (2013)

24.   A. J. Byrne, "Bioengineering and subjective approaches to the clinical evaluation of dry skin," *Int J Cosmet Sci* (2010)

25.   P. J. Caspers, G. W. Lucassen and G. J. Puppels, "Combined in vivo confocal Raman spectroscopy and confocal microscopy of human skin," *Biophys J* 85(1), 572-580 (2003)

26.   L. Chrit, P. Bastien, G. D. Sockalingum, D. Batisse, F. Leroy, M. Manfait and C. Hadjur, "An in vivo randomized study of human skin moisturization by a new confocal Raman fiber-optic microprobe: assessment of a glycerol-based hydration cream," *Skin Pharmacol Physiol* 19(4), 207-215 (2006)

27.   E. Ciampi, M. van Ginkel, P. J. McDonald, S. Pitts, E. Y. Bonnist, S. Singleton and A. M. Williamson, "Dynamic in vivo mapping of model moisturiser ingress into human skin by GARfield MRI," *NMR Biomed* 24(2), 135-144 (2011)

28.   M. Egawa and H. Tagami, "Comparison of the depth profiles of water and water-binding substances in the stratum corneum determined in vivo by Raman spectroscopy between the cheek and volar forearm skin: effects of age, seasonal changes and artificial forced hydration," *Br J Dermatol* 158(2), 251-260 (2008)

29.   M. Förster, M. A. Bolzinger, M. R. Rovere, O. Damour, G. Montagnac and S. Briançon, "Confocal Raman microspectroscopy for evaluating the stratum corneum removal by 3 standard methods," *Skin Pharmacol Physiol* 24(2), 103-112 (2011)

30.   N. Nakagawa, M. Matsumoto and S. Sakai, "In vivo measurement of the water content in the dermis by confocal Raman spectroscopy," *Skin Res Technol* 16(2), 137-141 (2010)

31.   R. Mateus, H. Abdalghafor, G. Oliveira, J. Hadgraft and M. E. Lane, "A new paradigm in dermatopharmacokinetics - Confocal Raman spectroscopy," *Int J Pharm* 444(1-2), 106-108 (2013)

32.   S. J. Jiang, J. Y. Chen, Z. F. Lu, J. Yao, D. F. Che and X. J. Zhou, "Biophysical and morphological changes in the stratum corneum lipids induced by UVB irradiation," *Journal of dermatological science* 44(1), 29-36 (2006)

33.   J. Chaiken, B. Deng, R. J. Bussjager, G. Shaheen, D. Rice, D. Stehlik and J. Fayos, "Instrument for near infrared emission spectroscopic probing of human fingertips in vivo," *Rev Sci Instrum* 81(3), 034301 (2010)

34.   A. Tfayli, O. Piot, S. Derancourt, G. Cadiot, M. D. Diebold, P. Bernard and M. Manfait, "In vivo analysis of tissue by Raman microprobe: examination of human skin lesions and esophagus Barrett's mucosa on an animal model," 609312-609312 (2006)

35.   A. Tfayli, E. Guillard, M. Manfait and A. Baillet-Guffroy, "Raman spectroscopy: feasibility of in vivo survey of stratum corneum lipids, effect of natural aging," *Eur J Dermatol* 22(1), 36-41 (2012)

36.   T. Bhattacharjee, P. Kumar, G. Maru, A. Ingle and C. M. Krishna, "Swiss bare mice: a suitable model for transcutaneous in vivo Raman spectroscopic studies of breast cancer," *Lasers Med Sci* (2013)

37.   J. Zhao, H. Lui, D. I. McLean and H. Zeng, "Automated autofluorescence background subtraction algorithm for biomedical Raman spectroscopy," *Appl Spectrosc* 61(11), 1225-1232 (2007)

38.   P. A. Philipsen, L. Knudsen, M. Gniadecka, M. H. Ravnbak and H. C. Wulf, "Diagnosis of malignant melanoma and basal cell carcinoma by in vivo NIR-FT Raman spectroscopy is independent of skin pigmentation," *Photochem Photobiol Sci* 12(5), 770-776 (2013)

39.   J. M. Amigo, C. Ravn, N. B. Gallagher and R. Bro, "A comparison of a common approach to partial least squares-discriminant analysis and classical least squares in hyperspectral imaging," *International Journal of Pharmaceutics* 373(1â€"2), 179-182 (2009)

40.   M. Pudlas, S. Koch, C. Bolwien, S. Thude, N. Jenne, T. Hirth, H. Walles and K. Schenke-Layland, "Raman spectroscopy: a noninvasive analysis tool for the discrimination of human skin cells," *Tissue Eng Part C Methods* 17(10), 1027-1040 (2011)

41.   S. Tfaili, C. Gobinet, G. Josse, J.-F. Angiboust, M. Manfait and O. Piot, "Confocal Raman microspectroscopy for skin characterization: a comparative study between human skin and pig skin," *Analyst* 137(16), 3673-3682 (2012)

42.   A. Savitzky and M. J. E. Golay, "Smoothing and differentiation of data by simplified least squares procedures," *Anal. Chem.* 36(1627-1639 (1964)

43.   A. Tfayli, O. Piot and M. Manfait, "Confocal Raman microspectroscopy on excised human skin: uncertainties in depth profiling and mathematical correction applied to dermatological drug permeation," *J Biophotonics* 1(2), 140-153 (2008)

44.   C. Merle, "Topologie de lipides cutanes : correlation avec le photo-vieillissement cutane et la penetration percutanee," in *Analytical Chemistry*, pp. 80-96, Paris Sud University, Chatenay-Malabry (2009).

45.   A. P. M. Lavrijsen, I. M. Higounenc, A. Weerheim, E. Oestmann, E. E. Tuinenburg, H. E. Bnddé and M. Ponec, "Validation of an in vivo extraction method for human stratum corneum ceramides," *Archives of Dermatological Research* 286(8), 495-503 (1994)

46.   C. Merle, C. Laugel, P. Chaminade and A. Baillet-Guffroy, "Quantitative study of the stratum corneum lipid classes by normal phase liquid chromatography: comparison between two universal detectors," *Journal of Liquid Chromatography & Related Technologies* 33(5), 629-644 (2010)

47.   J. Van Smeden, L. Hoppel, R. van der Heijden, T. Hankemeier, R. J. Vreeken and J. A. Bouwstra, "LC/MS analysis of stratum corneum lipids: ceramide profiling and discovery," *J Lipid Res* 52(6), 1211-1221 (2011)

48.   R. Bro, "Multiway calibration. Multilinear PLS," *Journal of Chemometrics* 10(1), 47-61 (1996)

49.   S. Wold, M. Sjostrom and L. Eriksson, "PLS-regression: a basic tool of chemometrics," *Chemometrics and Intelligent Laboratory Systems* 58(2), 109-130 (2001)

50.   I. Barman, C. R. Kong, N. C. Dingari, R. R. Dasari and M. S. Feld, "Development of robust calibration models using support vector machines for spectroscopic monitoring of blood glucose," *Anal Chem* 82(23), 9719-9726 (2010)

51.   S. Wold, H. Antti, F. Lindgren and J. Ã–hman, "Orthogonal signal correction of near-infrared spectra," *Chemometrics and Intelligent Laboratory Systems* 44(1â€"2), 175-185 (1998)

52.   H. T. Eastment and W. J. Krzanowski, "Cross-Validatory Choice of the Number of Components From a Principal Component Analysis," *Technometrics* 24(1), 73-77 (1982)

53.   G. Diana and C. Tommasi, "Cross-validation methods in principal component analysis: A comparison," *Statistical Methods and Applications* 11(1), 71-82 (2002)

54.   M. G. Ramirez-Elias, J. Alda and F. J. Gonzalez, "Noise and artifact characterization of in vivo Raman spectroscopy skin measurements," *Appl Spectrosc* 66(6), 650-655 (2012)

55.   J. Zhao, H. Lui, D. I. McLean and H. Zeng, "Integrated real-time Raman system for clinical in vivo skin analysis," *Skin Res Technol* 14(4), 484-492 (2008)

56.   M. Ponec, E. Boelsma, S. Gibbs and M. Mommaas, "Characterization of reconstructed skin models," *Skin Pharmacol Appl Skin Physiol* 15 Suppl 1(4-17 (2002)

57.   Y. Jokura, S. Ishikawa, H. Tokuda and G. Imokawa, "Molecular analysis of elastic properties of the stratum corneum by solid-state 13C-nuclear magnetic resonance spectroscopy," *J Invest Dermatol* 104(5), 806-812 (1995)

58.   P. J. Caspers, G. W. Lucassen and E. A. Carter, "In vivo confocal raman microspectroscopy of the skin: noninvasive determination of molecular concentration profiles," *J Invest Dermatol* 116(434-442 (2001)

59.   E. Guillard, A. Tfayli, M. Manfait and A. Baillet-Guffroy, "Thermal dependence of Raman descriptors of ceramides. Part II: effect of chains lengths and head group structures," *Anal Bioanal Chem* 399(3), 1201-1213 (2011)

60.   A. Tfayli, E. Guillard, M. Manfait and A. Baillet-Guffroy, "Thermal dependence of Raman descriptors of ceramides. Part I: effect of double bonds in hydrocarbon chains," *Anal Bioanal Chem* 397(3), 1281-1296 (2010)

61.   J. W. Fluhr, O. Kuss, T. Diepgen, S. Lazzerini, A. Pelosi, M. Gloor and E. Berardesca, "Testing for irritation with a multifactorial approach: comparison of eight non-invasive measuring techniques on five different irritation types," *British Journal of Dermatology* 145(5), 696-703 (2001)

62.   S. Verdier-Sevrain and F. Bonte, "Skin hydration: a review on its molecular mechanisms," *J Cosmet Dermatol* 6(2), 75-82 (2007)

63.   S. Bielfeldt, V. Schoder, U. Ely, A. Van Der Pol, J. De Sterke and K.-P. Wilhelm, "Assessment of human stratum corneum thickness and its barrier properties by in-vivo confocal Raman spectroscopy," *International Journal of Cosmetic Science* 31(6), 479-480 (2009)

64.   L. Norlen, I. Nicander, B. Lundh Rozell, S. Ollmar and B. Forslind, "Inter- and intra-individual differences in human stratum corneum lipid content related to physical parameters of skin barrier function in vivo," *J Invest Dermatol* 112(1), 72-77 (1999)

65.   S. L. Zhang, P. J. Caspers and G. J. Puppels, "In Vivo Confocal Raman Microspectroscopy of the Skin: Effect of Skin Care Products on Molecular Concentration Depth-Profiles," *Microscopy and Microanalysis* 11(SupplementS02), 790-791 (2005)

66.   M. Boncheva, J. de Sterke, P. J. Caspers and G. J. Puppels, "Depth profiling of Stratum corneum hydration in vivo: a comparison between conductance and confocal Raman spectroscopic measurements," *Exp Dermatol* 18(10), 870-876 (2009)

67.   M. Gniadecka, O. Faurskov Nielsen, D. H. Christensen and H. C. Wulf, "Structure of water, proteins, and lipids in intact human skin, hair, and nail," *J Invest Dermatol* 110(4), 393-398 (1998)

68.   M. Gniadecka, O. F. Nielsen, S. Wessel, M. Heidenheim, D. H. Christensen and H. C. Wulf, "Water and protein structure in photoaged and chronically aged skin," *J Invest Dermatol* 111(6), 1129-1133 (1998)

69.   G. E. Walrafen and Y. C. Chu, "Linearity between Structural Correlation Length and Correlated-Proton Raman Intensity from Amorphous Ice and Supercooled Water up to Dense Supercritical Steam," *The Journal of Physical Chemistry* 99(28), 11225-11229 (1995)

70.   P. Leary, F. Adar, R. Carlton, J. Reffner, F. Kang and R. Mueller, "Hydration Studies of Pharmaceuticals Using IR and Raman Spectroscopy," *Microscopy and Microanalysis* 13(SupplementS02), 1694-1695 (2007)

71.   J. Wu and T. G. Polefka, "Confocal Raman microspectroscopy of stratum corneum: a pre-clinical validation study," *Int J Cosmet Sci* 30(1), 47-56 (2008)

CONCLUSION GENERALE ET
PERSPECTIVES

Les travaux réalisés dans le cadre de cette thèse ont couvert différents aspects; partant du développement technique à l'identification des modifications structurales liées à la sécheresse cutanée, à la fois *ex vivo* et *in vivo,* ainsi qu'à la mise en évidences des mécanismes fonctionnels responsables des propriétés mécaniques du SC.

La première partie a consisté à établir des conditions d'analyse optimales pour la reproductibilité, la fiabilité et la robustesse des mesures. Dans ce cadre, le modèle simplifié de la chambre à humidité contrôlée RHC_2 nous a permis d'obtenir des valeurs d'HR stables, répétables et précises. Le contrôle d'HR autour du microscope associé à l'optimisation des paramètres d'acquisition pour l'analyse de la peau, a permis d'assurer la prise de mesures riches en informations dans des conditions bien déterminées.

Ensuite, ce travail de thèse décrit de nombreuses expériences, *ex vivo,* qui ont été réalisées sur le SC humain isolé.

La première étude expérimentale a consisté à l'établissement des descripteurs Raman de l'hydratation *ex vivo.* Malgré l'abondance des données bibliographiques reposant sur des travaux en spectroscopie Raman pour l'étude de l'état d'hydratation cutanée, ce travail a eu pour objectif de poser un regard critique sur la lecture spectrale de l'hydratation cutanée tout en mettant l'accent sur l'importance des données quantitatives. Ces nouvelles analyses *ex vivo* ont montré que l'ultra-structure moléculaire de la peau est influencée par les conditions environnementales.

Nous avons remis à jour la notion d'eau liée et non-liée au SC, et introduit l'effet de la structure de l'eau sur le mécanisme moléculaire du SC. Ainsi, nous avons apporté de nouvelles données permettant de démontrer que l'eau partiellement liée peu considérée dans les études cutanées, est un élément majeur dans l'équilibre conformationnel et structural des lipides et protéines.

En plus des aspects moléculaires, l'état mécanique de la peau est un des paramètres les plus importants dans le cas de problèmes de sécheresse. . Nous nous sommes intéressés aux mécanismes moléculaires impliqués dans le stress mécanique du SC ; en se servant de l'outil Raman pour la caractérisation des modifications structurales.

222

Dans un premier temps, nos travaux ont permis d'associer un état mécanique à son état d'hydratation et à l'organisation supramoléculaire des composants du SC. Il a donc été démontré qu'un équilibre entre les hélices α et les feuillets β est à la fois associée aux changements de la structure de l'eau et l'évolution du stress mécanique. Il s'agit d'une première description succincte des caractéristiques supramoléculaires de la peau sèche et leur lien avec le stress mécanique.

Dans un second temps, l'exploitation de l'empreinte vibrationnelle a été approfondie au-delà de l'aspect moléculaire par une corrélation directe entre les spectres Raman et la traction uni-axiale du SC. La régression partielle par moindres carrés (PLS) appliquée aux signatures Raman et aux pourcentages de traction a permis d'établir une bonne corrélation et de proposer un modèle de prédiction. Ce dernier permet à travers la seule signature Raman de prédire le pourcentage d'allongement du SC. Ce travail a été le premier à investiguer une approche mécanique via les spectres Raman sur les tissus cutanés ouvrant ainsi une nouvelle perspective vers une compréhension plus globale de la peau.

L'objectif global de la recherche sur la sécheresse cutanée étant toujours de rétablir un état d'hydratation « normal » ; nous nous sommes intéressés à étudier les effets de produits « hydratants ». Sur la base des descripteurs déjà établis lors des études précédentes, nous nous sommes intéressés à la compréhension des modes d'action de différents produits. Une étude comparative a donc permis d'introduire une nouvelle vision dans l'évaluation des effets des produits hydratants. En plus des propriétés hygroscopiques et de rétention d'eau, l'évolution de l'organisation des lipides et la pénétration des produits a conduit à une classification de ces produits dans les: occlusifs, modulateurs de la barrière lipidique, humectants et émollients, mais en s'appuyant sur des observations à l'échelle moléculaire.

Dans la logique de notre démarche, allant du moléculaire vers le vivant, la dernière partie de cette thèse consistait à la transposition des analyses vers l'*in vivo*. Dans ce cadre, les mesures Raman *in vivo* ont été réalisées sur un panel de patients sains en parallèle avec un ensemble d'analyses biométriques (pH, PIE, cornéométrie) et une quantification relative des classes lipidiques. L'originalité de notre démarche résidait non-seulement dans l'association des différents paramètres afin de décrire le statut de la peau, mais surtout dans la création d'un modèle mathématique permettant d'obtenir ces paramètres à partir de la signature Raman. Un spectre Raman permet ainsi d'évaluer les pourcentages des différentes classes lipidiques, le pH, la PIE, l'hydratation globale du SC, y compris la structure de l'eau, des protéines et des lipides. Le spectre Raman se présente ainsi comme un véritable « QR code de la peau ». Cette première investigation du QR code de la peau nous a fourni des résultats très prometteurs. Ces résultats sont à consolider en élargissant la bibliothèque spectrale sur des patients présentant des pathologies cutanées.

D'une façon plus générale, ce travail de thèse a démontré l'intérêt de la micro spectroscopie Raman en tant que technique émergente présentant des atouts par rapport aux autres approches existantes pour caractériser globalement la peau.

RÉFÉRENCES

[1] M. Egawa, T. Hirao, M. Takahashi, In vivo estimation of stratum corneum thickness from water concentration profiles obtained with Raman spectroscopy, Acta Derm Venereol, 87 (2007) 4-8.

[2] J.M. Crowther, A. Sieg, P. Blenkiron, C. Marcott, P.J. Matts, J.R. Kaczvinsky, A.V. Rawlings, Measuring the effects of topical moisturizers on changes in stratum corneum thickness, water gradients and hydration in vivo, Br J Dermatol, 159 (2008) 567-577.

[3] K. Levi, R.J. Weber, J.Q. Do, R.H. Dauskardt, Drying stress and damage processes in human stratum corneum, Int J Cosmet Sci, 32 (2010) 276-293.

[4] M. Egawa, H. Tagami, Comparison of the depth profiles of water and water-binding substances in the stratum corneum determined in vivo by Raman spectroscopy between the cheek and volar forearm skin: effects of age, seasonal changes and artificial forced hydration, Br J Dermatol, 158 (2008) 251-260.

[5] P.J. Caspers, G.W. Lucassen, E.A. Carter, In vivo confocal raman microspectroscopy of the skin: noninvasive determination of molecular concentration profiles, J Invest Dermatol, 116 (2001) 434-442.

[6] E. Christophers, A.M. Kligman, Visualization of the Cell Layers of the Stratum Corneum1, The Journal of Investigative Dermatology, 42 (1964) 407-409.

[7] O. Simonetti, A.J. Hoogstraate, W. Bialik, J.A. Kempenaar, A.H. Schrijvers, H.E. Bodde, M. Ponec, Visualization of diffusion pathways across the stratum corneum of native and in-vitro-reconstructed epidermis by confocal laser scanning microscopy, Arch Dermatol Res, 287 (1995) 465-473.

[8] A.S. Michaels, S.K. Chandrasekaran, J.E. Shaw, Drug permeation through human skin: Theory and invitro experimental measurement, AIChE Journal, 21 (1975) 985-996.

[9] P. Garidel, B. Fölting, I. Schaller, A. Kerth, The microstructure of the stratum corneum lipid barrier: Mid-infrared spectroscopic studies of hydrated ceramide:palmitic acid:cholesterol model systems, Biophysical Chemistry, 150 (2010) 144-156.

[10] R.R. Wickett, M.O. Visscher, Structure and function of the epidermalÂ barrier, American journal of infection control, 34 (2006) S98-S110.

[11] I.R. Scott, C.R. Harding, Filaggrin breakdown to water binding compounds during development of the rat stratum corneum is controlled by the water activity of the environment, Dev Biol, 115 (1986) 84-92.

[12] G. Zhang, D.J. Moore, R. Mendelsohn, C.R. Flach, Vibrational microspectroscopy and imaging of molecular composition and structure during human corneocyte maturation, J Invest Dermatol, 126 (2006) 1088-1094.

[13] M. Simon, D. Bernard, A.M. Minondo, C. Camus, F. Fiat, P. Corcuff, R. Schmidt, G. Serre, Persistence of both peripheral and non-peripheral corneodesmosomes in the upper stratum corneum of winter xerosis skin versus only peripheral in normal skin, J Invest Dermatol, 116 (2001) 23-30.

[14] L. Norlen, I. Nicander, B. Lundh Rozell, S. Ollmar, B. Forslind, Inter- and intra-individual differences in human stratum corneum lipid content related to physical parameters of skin barrier function in vivo, J Invest Dermatol, 112 (1999) 72-77.

[15] A. Baroni, E. Buommino, V. De Gregorio, E. Ruocco, V. Ruocco, R. Wolf, Structure and function of the epidermis related to barrier properties, Clinics in dermatology, 30 (2012) 257-262.

[16] L. Norlen, I. Plasencia, A. Simonsen, Human stratum corneum lipid organization as observed by atomic force microscopy on Langmuir-Blodgett films, J Struct Biol, 158 (2007) 386-400.

[17] S. Pappinen, M. Hermansson, J. Kuntsche, P. Somerharju, P. Wertz, A. Urtti, M. Suhonen, Comparison of rat epidermal keratinocyte organotypic culture (ROC) with intact human skin: lipid composition and thermal phase behavior of the stratum corneum, Biochim Biophys Acta, 1778 (2008) 824-834.

[18] J. Van Smeden, L. Hoppel, R. van der Heijden, T. Hankemeier, R.J. Vreeken, J.A. Bouwstra, LC/MS analysis of stratum corneum lipids: ceramide profiling and discovery, J Lipid Res, 52 (2011) 1211-1221.

[19] K.C. Madison, D.C. Swartzendruber, P.W. Wertz, Presence of intact intercellular lamellae in the upper layers of the stratum corneum, J Invest Dermatol, 88 (1987) 714-718.

[20] Y. Mizutani, S. Mitsutake, K. Tsuji, A. Kihara, Y. Igarashi, Ceramide biosynthesis in keratinocyte and its role in skin function, Biochimie, 91 (2009) 784-790.

[21] M. Engelke, J.M. Jensen, S. Ekanayake-Mudiyanselage, E. Proksch, Effects of xerosis and ageing on epidermal proliferation and differentiation, Br J Dermatol, 137 (1997) 219-225.

[22] C. Das, P.D. Olmsted, M. Noro, Simulation studies of stratum corneum lipid mixtures, Biophys J, 97 (2009) 1941-1951.

[23] J.A. Bouwstra, G.S. Gooris, M. Ponec, Skin lipid organization, composition and barrier function, International Journal of Cosmetic Science, 30 (2008) 388-388.

[24] M. Behne, Y. Uchida, T. Seki, P.O. de Montellano, P.M. Elias, W.M. Holleran, Omega-Hydroxyceramides are Required for Corneocyte Lipid Envelope (CLE) Formation and Normal Epidermal Permeability Barrier Function, J Investig Dermatol, 114 (2000) 185-192.

[25] Y. Zheng, H. Yin, W.E. Boeglin, P.M. Elias, D. Crumrine, D.R. Beier, A.R. Brash, Lipoxygenases mediate the effect of essential fatty acid in skin barrier formation: a proposed role in releasing omega-hydroxyceramide for construction of the corneocyte lipid envelope, J Biol Chem, 286 (2011) 24046-24056.

[26] A. Oren, T. Ganz, L. Liu, T. Meerloo, In human epidermis, beta-defensin 2 is packaged in lamellar bodies, Experimental and Molecular Pathology, 74 (2003) 180-182.

[27] Braff, H. Marissa, N. Di, Anna, Gallo, L. Richard, Keratinocytes store the antimicrobial peptide cathelicidin in lamellar bodies, Nature Publishing Group, New York, NY, ETATS-UNIS, 2005.

[28] D. Mohammed, P.J. Matts, J. Hadgraft, M.E. Lane, Influence of Aqueous Cream BP on corneocyte size, maturity, skin protease activity, protein content and transepidermal water loss, Br J Dermatol, 164 (2011) 1304-1310.

[29] P.M. Elias, Epidermal barrier function: intercellular lamellar lipid structures, origin, composition and metabolism, Journal of Controlled Release, 15 (1991) 199-208.

[30] R.B. Presland, M.K. Kuechle, S.P. Lewis, P. Fleckman, B.A. Dale, Regulated Expression of Human Filaggrin in Keratinocytes Results in Cytoskeletal Disruption, Loss of Cellâ€"Cell Adhesion, and Cell Cycle Arrest, Experimental Cell Research, 270 (2001) 199-213.

[31] A. Sandilands, C. Sutherland, A.D. Irvine, W.H. McLean, Filaggrin in the frontline: role in skin barrier function and disease, J Cell Sci, 122 (2009) 1285-1294.

[32] J.W. Fluhr, R. Darlenski, C. Surber, Glycerol and the skin: holistic approach to its origin and functions, Br J Dermatol, 159 (2008) 23-34.

[33] A.V. Rawlings, I.R. Scott, C.R. Harding, P.A. Bowser, Stratum corneum moisturization at the molecular level, J Invest Dermatol, 103 (1994) 731-741.

[34] G. Denecker, P. Ovaere, P. Vandenabeele, W. Declercq, Caspase-14 reveals its secrets, J Cell Biol, 180 (2008) 451-458.

[35] W.M. Holleran, Y. Takagi, Y. Uchida, Epidermal sphingolipids: Metabolism, function, and roles in skin disorders, FEBS Letters, 580 (2006) 5456-5466.

[36] A. Pons-Guiraud, Dry skin in dermatology: a complex physiopathology, Journal of the European Academy of Dermatology and Venereology : JEADV, 21 Suppl 2 (2007) 1-4.

[37] A.V. Rawlings, P.J. Matts, Stratum corneum moisturization at the molecular level: an update in relation to the dry skin cycle, J Invest Dermatol, 124 (2005) 1099-1110.

[38] J.A. Bouwstra, M. Ponec, The skin barrier in healthy and diseased state, Biochim Biophys Acta, 1758 (2006) 2080-2095.

[39] E. Simpson, A. Bohling, S. Bielfeldt, C. Bosc, N. Kerrouche, Improvement of skin barrier function in atopic dermatitis patients with a new moisturizer containing a ceramide precursor, J Dermatolog Treat, 24 (2013) 122-125.

[40] A.V. Rawlings, Trends in stratum corneum research and the management of dry skin conditions, Int J Cosmet Sci, 25 (2003) 63-95.

[41] A.V. Rawlings, C.R. Harding, Moisturization and skin barrier function, Dermatol Ther, 17 Suppl 1 (2004) 43-48.

[42] E. Alanen, J. Nuutinen, K. Nicklén, T. Lahtinen, J. Mönkkönen, Measurement of hydration in the stratum corneum with the MoistureMeter and comparison with the Corneometer, Skin Research and Technology, 10 (2004) 32-37.

[43] T.-C. Chou, K.-H. Lin, S.-M. Wang, C.-W. Lee, S.-B. Su, T.-S. Shih, H.-Y. Chang, Transepidermal water loss and skin capacitance alterations among workers in an ultra-low humidity environment, Archives of Dermatological Research, 296 (2005) 489-495.

[44] J.W. Fluhr, M. Gloor, S. Lazzerini, P. Kleesz, R. Grieshaber, E. Berardesca, Comparative study of five instruments measuring stratum corneum hydration (Corneometer CM 820 and CM 825, Skicon 200, Nova DPM 9003, DermaLab). Part II. In vivo, Skin Research and Technology, 5 (1999) 171-178.

[45] D. Hildebrandt, K. Ziegler, U. Wollina, Electrical impedance and transepidermal water loss of healthy human skin under different conditions, Skin Research and Technology, 4 (1998) 130-134.

[46] M. Boncheva, J. de Sterke, P.J. Caspers, G.J. Puppels, Depth profiling of Stratum corneum hydration in vivo: a comparison between conductance and confocal Raman spectroscopic measurements, Exp Dermatol, 18 (2009) 870-876.

[47] L. Chrit, P. Bastien, B. Biatry, J.T. Simonnet, A. Potter, A.M. Minondo, F. Flament, R. Bazin, G.D. Sockalingum, F. Leroy, M. Manfait, C. Hadjur, In vitro and in vivo confocal Raman study of human skin hydration: assessment of a new moisturizing agent, pMPC, Biopolymers, 85 (2007) 359-369.

[48] L. Chrit, P. Bastien, G.D. Sockalingum, D. Batisse, F. Leroy, M. Manfait, C. Hadjur, An in vivo randomized study of human skin moisturization by a new confocal Raman fiber-optic microprobe: assessment of a glycerol-based hydration cream, Skin Pharmacol Physiol, 19 (2006) 207-215.

[49] N. Nakagawa, M. Matsumoto, S. Sakai, In vivo measurement of the water content in the dermis by confocal Raman spectroscopy, Skin Res Technol, 16 (2010) 137-141.

[50] P.J. Caspers, G.W. Lucassen, E.A. Carter, H.A. Bruining, G.J. Puppels, In vivo confocal Raman microspectroscopy of the skin: noninvasive determination of molecular concentration profiles, J Invest Dermatol, 116 (2001) 434-442.

[51] T. Gambichler, J. Huyn, N.S. Tomi, G. Moussa, C. Moll, A. Sommer, P. Altmeyer, K. Hoffmann, A comparative pilot study on ultraviolet-induced skin changes assessed by noninvasive imaging techniques in vivo, Photochem Photobiol, 82 (2006) 1103-1107.

[52] M. Mogensen, H.A. Morsy, L. Thrane, G.B. Jemec, Morphology and epidermal thickness of normal skin imaged by optical coherence tomography, Dermatology, 217 (2008) 14-20.

[53] A.V. Rawlings, P.J. Matts, C.D. Anderson, M.S. Roberts, Skin biology, xerosis, barrier repair and measurement, Drug Discovery Today: Disease Mechanisms, 5 (2008) e127-e136.

[54] C. Rosado, P. Pinto, L.M. Rodrigues, Comparative assessment of the performance of two generations of Tewameter: TM210 and TM300, Int J Cosmet Sci, 27 (2005) 237-241.

[55] R. Darlenski, S. Sassning, N. Tsankov, J.W. Fluhr, Non-invasive in vivo methods for investigation of the skin barrier physical properties, Eur J Pharm Biopharm, 72 (2009) 295-303.

[56] Chan, W. James, Taylor, S. Douglas, Zwerdling, Theodore, Lane, M. Stephen, Lhara, Ko, Huser, Thomas, Micro-Raman spectroscopy detects individual neoplastic and normal hematopoietic cells, Cell Press, Cambridge, MA, ETATS-UNIS, 2005.

[57] Rosado, Catarina, Pinto, Pedro, R. Monteiro, Luis, Modeling TEWM-desorption curves: a new practical approach for the quantitative in vivo assessment of skin barrier, Blackwell, Oxford, ROYAUME-UNI, 2005.

[58] L. Rodrigues, EEMCO Guidance to the in vivo Assessment of Tensile Functional Properties of the Skin, Skin Pharmacology and Physiology, 14 (2001) 52-67.

[59] P. Clarys, R. Clijsen, J. Taeymans, A.O. Barel, Hydration measurements of the stratum corneum: comparison between the capacitance method (digital version of the Corneometer CM 825®) and the impedance method (Skicon-200EX®), Skin Research and Technology, 18 316-323.

[60] T. Rosendal, Concluding Studies on the Conducting Properties of Human Skin to Alternating Current, Acta Physiologica Scandinavica, 9 (1945) 39-49.

[61] E. Berardesca, Bioengineering of the Skin: Methods and Instrumentation, CRC Press, 1995.

[62] S. Bielfeldt, V. Schoder, U. Ely, A. Van Der Pol, J. De Sterke, K.-P. Wilhelm, Assessment of human stratum corneum thickness and its barrier properties by in-vivo confocal Raman spectroscopy, International Journal of Cosmetic Science, 31 (2009) 479-480.

[63] S.L. Zhang, P.J. Caspers, G.J. Puppels, In Vivo Confocal Raman Microspectroscopy of the Skin: Effect of Skin Care Products on Molecular Concentration Depth-Profiles, Microscopy and Microanalysis, 11 (2005) 790-791.

[64] A.H.P.I. Barth, Biological and biomedical infrared spectroscopy, in, IOS Press, Amsterdam; Washington, DC, 2009.

[65] M. Egawa, In vivo simultaneous measurement of urea and water in the human stratum corneum by diffuse-reflectance near-infrared spectroscopy, Skin Res Technol, 15 (2009) 195-199.

[66] J. Wu, T.G. Polefka, Confocal Raman microspectroscopy of stratum corneum: a pre-clinical validation study, International Journal of Cosmetic Science, 30 (2008) 47-56.

[67] A. Tfayli, E. Guillard, M. Manfait, A. Baillet-Guffroy, Thermal dependence of Raman descriptors of ceramides. Part I: effect of double bonds in hydrocarbon chains, Anal Bioanal Chem, 397 (2010) 1281-1296.

[68] S. Raudenkolb, S. Wartewig, G. Brezesinski, S.S. Funari, R.H. Neubert, Hydration properties of N-(alpha-hydroxyacyl)-sphingosine: X-ray powder diffraction and FT-Raman spectroscopic studies, Chem Phys Lipids, 136 (2005) 13-22.

[69] E. Guillard, A. Tfayli, M. Manfait, A. Baillet-Guffroy, Thermal dependence of Raman descriptors of ceramides. Part II: effect of chains lengths and head group structures, Anal Bioanal Chem, 399 (2011) 1201-1213.

[70] J. Caussin, G.S. Gooris, J.A. Bouwstra, FTIR studies show lipophilic moisturizers to interact with stratum corneum lipids, rendering the more densely packed, Biochim Biophys Acta, 1778 (2008) 1517-1524.

[71] E.C. Guillard, C. Laugel, A. Baillet-Guffroy, Molecular interactions of penetration enhancers within ceramides organization: a FTIR approach, Eur J Pharm Sci, 36 (2009) 192-199.

[72] A. Tfayli, O. Piot, F. Draux, F. Pitre, M. Manfait, Molecular characterization of reconstructed skin model by Raman microspectroscopy: comparison with excised human skin, Biopolymers, 87 (2007) 261-274.

[73] A. Tfayli, E. Guillard, M. Manfait, A. Baillet-Guffroy, Raman spectroscopy: feasibility of in vivo survey of stratum corneum lipids, effect of natural aging, Eur J Dermatol, 22 (2012) 36-41.

[74] M. Gniadecka, O.F. Nielsen, S. Wessel, M. Heidenheim, D.H. Christensen, H.C. Wulf, Water and protein structure in photoaged and chronically aged skin, J Invest Dermatol, 111 (1998) 1129-1133.

[75] G.L. Wilkes, I.A. Brown, R.H. Wildnauer, The biomechanical properties of skin, CRC Crit Rev Bioeng, 1 (1973) 453-495.

[76] Marks, R, The stratum corneum barrier: The final frontier, American Society for Nutrition, Bethesda, MD, ETATS-UNIS, 2004.

[77] Y. Yuan, R. Verma, Measuring microelastic properties of stratum corneum, Colloids and Surfaces B: Biointerfaces, 48 (2006) 6-12.

[78] C. Pailler-Mattei, S. Pavan, R. Vargiolu, F. Pirot, F. Falson, H. Zahouani, Contribution of stratum corneum in determining bio-tribological properties of the human skin, Wear, 263 (2007) 1038-1043.

[79] R.H. Wildnauer, J.W. Bothwell, Douglass, Stratum corneum biomechanical properties. I. Influence of relative humidity on normal and extracted human stratum corneum, J Invest Dermatol, 56 (1971) 72-78.

[80] P. Agache, J.P. Boyer, R. Laurent, Biomechanical properties and microscopic morphology of human stratum corneum incubated on a wet pad in vitro, Arch Dermatol Forsch, 246 (1973) 271-283.

[81] Y.S. Papir, K.H. Hsu, R.H. Wildnauer, The mechanical properties of stratum corneum. I. The effect of water and ambient temperature on the tensile properties of newborn rat stratum corneum, Biochim Biophys Acta, 399 (1975) 170-180.

[82] A. Delalleau, G. Josse, J.M. Lagarde, H. Zahouani, J.M. Bergheau, A nonlinear elastic behavior to identify the mechanical parameters of human skin in vivo, Skin Research and Technology, 14 (2008) 152-164.

[83] A. Delalleau, G. Josse, J.-M. Lagarde, H. Zahouani, J.-M. Bergheau, Characterization of the mechanical properties of skin by inverse analysis combined with the indentation test, Journal of biomechanics, 39 (2006) 1603-1610.

[84] A.C. PARK, C.B. BADDIEL, Rheology of stratum corneum-I: A molecular interpretation of the stress-strain curve, J. Soc. Cosmet. Chem., 23 (1972) 3-12.

[85] K.S. Wu, W.W. van Osdol, R.H. Dauskardt, Mechanical properties of human stratum corneum: effects of temperature, hydration, and chemical treatment, Biomaterials, 27 (2006) 785-795.

[86] C.S. Nicolopoulos, P.V. Giannoudis, K.D. Glaros, J.C. Barbenel, In vitro study of the failure of skin surface after influence of hydration and preconditioning, Arch Dermatol Res, 290 (1998) 638-640.

[87] B.F. Van Duzee, The influence of water content, chemical treatment and temperature on the rheological properties of stratum corneum, J Invest Dermatol, 71 (1978) 140-144.

[88] M.A. Wolfram, N.F. Wolejsza, K. Laden, Biomechanical properties of delipidized stratum corneum, J Invest Dermatol, 59 (1972) 421-426.

[89] K. Levi, R.H. Dauskardt, Application of substrate curvature method to differentiate drying stresses in topical coatings and human stratum corneum, Int J Cosmet Sci, 32 (2010) 294-298.

[90] D.S. Gardner, P.A. Flinn, Mechanical stress as a function of temperature in aluminum films, Electron Devices, IEEE Transactions on, 35 (1988) 2160-2169.

[91] M. Moske, K. Samwer, New UHV dilatometer for precise measurement of internal stresses in thin binary‐alloy films from 20 to 750 K, Review of Scientific Instruments, 59 (1988) 2012-2017.

[92] S. Bader, E.M. Kalaugher, E. Arzt, Comparison of mechanical properties and microstructure of Al(1 wt.%Si) and Al(1 wt.%Si, 0.5 wt.%Cu) thin films, Thin Solid Films, 263 (1995) 175-184.

[93] K. Levi, A. Kwan, A.S. Rhines, M. Gorcea, D.J. Moore, R.H. Dauskardt, Emollient molecule effects on the drying stresses in human stratum corneum, Br J Dermatol, 163 (2010) 695-703.

[94] K. Levi, A. Kwan, A.S. Rhines, M. Gorcea, D.J. Moore, R.H. Dauskardt, Effect of glycerin on drying stresses in human stratum corneum, J Dermatol Sci, 61 (2011) 129-131.

[95] M. Gasior-Glogowska, M. Komorowska, J. Hanuza, M. Maczka, A. Zajac, M. Ptak, R. Bedzinski, M. Kobielarz, K. Maksymowicz, P. Kuropka, S. Szotek, FT-Raman spectroscopic study of human skin subjected to uniaxial stress, Journal of the Mechanical Behavior of Biomedical Materials, 18 (2013) 240-252.

[96] B.A. Roeder, K. Kokini, J.E. Sturgis, J.P. Robinson, S.L. Voytik-Harbin, Tensile mechanical properties of three-dimensional type I collagen extracellular matrices with varied microstructure, J Biomech Eng, 124 (2002) 214-222.

[97] J.G. Snedeker, P. Niederer, F.R. Schmidlin, M. Farshad, C.K. Demetropoulos, J.B. Lee, K.H. Yang, Strain-rate dependent material properties of the porcine and human kidney capsule, Journal of biomechanics, 38 (2005) 1011-1021.

[98] P. Agache, Physiologie de la peau et explorations fonctionnelles cutanées, Edition Medicale International, 2000.

[99] P.A. Payne, Measurement of properties and function of skin, Clin Phys Physiol Meas, 12 (1991) 105-129.

[100] S. Diridollou, M. Berson, V. Vabre, D. Black, B. Karlsson, F. Auriol, J.M. Gregoire, C. Yvon, L. Vaillant, Y. Gall, F. Patat, An In Vivo Method for Measuring the Mechanical Properties of the Skin Using Ultrasound, Ultrasound in medicine & biology, 24 (1998) 215-224.

[101] F.M. Hendriks, D. Brokken, J.T.W.M. Van Eemeren, C.W.J. Oomens, F.P.T. Baaijens, J.B.A.M. Horsten, A numerical-experimental method to characterize the non-linear mechanical behaviour of human skin, Skin Research and Technology, 9 (2003) 274-283.

[102] N.B. Colthup, L.H. Daly, S.E. Wiberley, CHAPTER 14 - THE THEORETICAL ANALYSIS OF MOLECULAR VIBRATIONS, in: Introduction to Infrared and Raman Spectroscopy (Third Edition), Academic Press, San Diego, 1990, pp. 483-542.

[103] M. Diem, Introduction to modern vibrational spectroscopy, Wiley, 1993.

[104] J.R. Ferraro, K. Nakamoto, C.W. Brown, Introductory Raman Spectroscopy, Academic Press, 2003.

[105] L.A. Woodward, Introduction to the theory of molecular vibrations and vibrational spectroscopy, Clarendon Press, 1972.

[106] N.B. Colthup, Introduction to infrared and Raman spectroscopy, Academic Press, New York, 1964.

[107] T. W.J.O, Basic laser Raman spectroscopy : Jack Loader, Heyden & Son Ltd., London, 1970, pp. 112, price Â£1.90, Journal of Molecular Structure, 9 (1971) 490.

[108] M.C. Tobin, Laser Raman spectroscopy, Wiley-Interscience, New York, 1971.

[109] D.A. Long, Raman spectroscopy, McGraw-Hill, New York, 1977.

[110] I. Wiley, Journal of Raman spectroscopy, in, John Wiley & Sons, Chichester; New York, NY.

[111] E.D.G. Smith, Modern Raman spectroscopy : a practical approach, J. Wiley, Hoboken, NJ, 2005.

[112] R.L. McCreery, Raman spectroscopy for chemical analysis, John Wiley & Sons, New York, 2000.

[113] F. Adar, Analytical Vibrational Spectroscopy - NIR, IR, and Raman, Spectroscopy, 26 (2011) 14-23.

[114] A. Michelet, M. Boiret, F. Lemhachheche, L. Malec, A. Tfayli, E. Ziemons, Utilisation de la spectrométrie Raman dans le domaine pharmaceutique, Français, 23 (2013) 97-117.

[115] P.J. Caspers, G.W. Lucassen, R. Wolthuis, H.A. Bruining, G.J. Puppels, In vitro and in vivo Raman spectroscopy of human skin, Biospectroscopy, 4 (1998) S31-39.

[116] A. Savitzky, M.J.E. Golay, Smoothing and differentiation of data by simplified least squares procedures, Anal. Chem., 36 (1964) 1627-1639.

[117] J. Zhao, H. Lui, D.I. McLean, H. Zeng, Automated autofluorescence background subtraction algorithm for biomedical Raman spectroscopy, Appl Spectrosc, 61 (2007) 1225-1232.

[118] W.O.M.P.S.M.D.J.A. George, Infrared spectroscopy, Published on behalf of ACOL, London, by Wiley, Chichester [West Sussex]; New York, 1987.

[119] R.T. Conley, Infrared spectroscopy, Allyn and Bacon, Boston, 1966.

[120] B.A.D.J. Stuart, Modern infrared spectroscopy, Published on behalf of ACOL (University of Greenwich) by Wiley, New York, 1996.

[121] B. Stuart, Infrared spectroscopy fundamentals and applications, in, J. Wiley, Chichester, Eng.; Hoboken, N.J., 2004.

[122] D.N. Kendall, Applied infrared spectroscopy, Reinhold Pub. Corp., New York, 1966.

[123] J. Withrow, Infrared spectroscopy, in, Research World, Delhi, 2012.

[124] A.L. Smith, Applied infrared spectroscopy : fundamentals, techniques, and analytical problem-solving, Wiley, New York, 1979.

[125] B.C. Smith, Fundamentals of Fourier transform infrared spectroscopy, CRC Press, Boca Raton, 1996.

[126] R.G.J.S.B.C.A.D.M. Miller, Laboratory methods in infrared spectroscopy, Heyden, London; New York, 1972.

[127] M. Milosevic, Internal reflection and ATR spectroscopy, in, John Wiley & Sons, Hoboken, N.J., 2012.

[128] A.A.L.A.H.M.S.A.S.R.N.A.K.S.T.S.S.T.H. Kandhro, Application of attenuated total reflectance Fourier transform infrared spectroscopy for determination of cefixime in oral pharmaceutical formulations, SAA Spectrochimica Acta Part A: Molecular and Biomolecular Spectroscopy, 115 (2013) 51-56.

[129] E.J.K.R. Elzinga, In situ ATR-FTIR spectroscopic analysis of the co-adsorption of orthophosphate and Cd(II) onto hematite, GCA Geochimica et Cosmochimica Acta, 117 (2013) 53-64.

[130] J.M. Chalmers, P.R. Griffiths, Handbook of vibrational spectroscopy, John Wiley & Sons, London, 2002.

[131] G.G. Stoney, The tension of metallic films deposited by electrolysis, in: Proceedings of the Royal Society of London Series B, Biological sciences, vol. 82, 1909, pp. 172.

[132] H. Shinzawa, K. Awa, W. Kanematsu, Y. Ozaki, Multivariate data analysis for Raman spectroscopic imaging, Journal of Raman Spectroscopy, 40 (2009) 1720-1725.

[133] J.F. Hair, Multivariate Data Analysis, Prentice-Hall International, 1998.

[134] J.F. Hair, R.E. Anderson, Multivariate data analysis, Prentice Hall Higher Education, 2010.

[135] D. Cozzolino, W.U. Cynkar, N. Shah, P. Smith, Multivariate data analysis applied to spectroscopy: Potential application to juice and fruit quality, Food Research International, 44 (2011) 1888-1896.

[136] M.L. O'Connell, A.G. Ryder, M.N. Leger, T. Howley, Qualitative analysis using Raman spectroscopy and chemometrics: a comprehensive model system for narcotics analysis, Appl Spectrosc, 64 (2010) 1109-1121.

[137] W.G.D.W.J.S.D.R.U.S.E.P.A. Glen, Principal components analysis and partial least squares regression, U.S. Environmental Protection Agency, Washington, D.C.?, 1992.

[138] I.T. Jolliffe, Principal component analysis, Springer-Verlag, New York, 1986.

[139] G.H. Dunteman, Principal components analysis, Sage Publications, Newbury Park, 1989.

[140] M. Castanys, M.J. Soneira, R. Perez-Pueyo, Automatic Identification of Artistic Pigments by Raman Spectroscopy Using Fuzzy Logic and Principal Component Analysis, Laser Chemistry, 2006.

[141] R.Y. Sato-Berru, E.V. Mejia-Uriarte, C. Frausto-Reyes, M. Villagran-Muniz, H.M. S, J.M. Saniger, Application of principal component analysis and Raman spectroscopy in the analysis of polycrystalline BaTiO3 at high pressure, Spectrochimica Acta Part A: Molecular and Biomolecular Spectroscopy, 66 (2007) 557-560.

[142] C. Syms, Principal Components Analysis, in: Encyclopedia of Ecology, Academic Press, Oxford, 2008, pp. 2940-2949.

[143] K.H. Esbensen, P. Geladi, 2.13 - Principal Component Analysis: Concept, Geometrical Interpretation, Mathematical Background, Algorithms, History, Practice, in: Comprehensive Chemometrics, vol. 2, Elsevier, Oxford, 2009, pp. 211-226.

[144] J.-B. Lohmöller, Introduction to PLS estimation of path models with latent variables, including some recent developments on mixed scales variables, Neubiberg, 1982.

[145] H. Wold, Systems analysis by partial least squares (PLS), Laxenburg, 1983.

[146] J.-B.W.H. Lohmöller, Three-mode path models with latent variables and partial least squares (PLS) parameter estimation, in, Neubiberg, 1980.

[147] R.W. Gerlach, B.R. Kowalski, H.O.A. Wold, Partial least-squares path modelling with latent variables, Analytica Chimica Acta, 112 (1979) 417-421.

[148] P.N.Q.D. Weinmann, Mise au point d'une methode d'analyse quantitative du cholesterol et de ses esters dans les lesions d'atherosclerose humaine par spectrometrie de diffusion raman dans le proche infrarouge, in: Sciences Medicales : ECP, [s.n.], [S.l.], 1996.

[149] V. Esposito Vinzi, Handbook of partial least squares concepts, methods and applications, in, Springer, Berlin; New York, 2010.

[150] S.C.C.E.P.J.v. Serneels, Influence properties of partial least squares regression, Katholieke Universiteit Leuven, [Faculteit Economische en Toegepaste Economische Wetenschappen], Departement Toegepaste Economische Wetenschappen, Leuven, 2003.

[151] S. Wold, M. SjÃ¶strÃ¶m, L. Eriksson, PLS-regression: a basic tool of chemometrics, Chemometrics and Intelligent Laboratory Systems, 58 (2001) 109-130.

[152] J.N.M.J.C. Miller, Statistics and chemometrics for analytical chemistry, Prentice Hall, Harlow, 2010.

[153] D. Cebeci Maltas, K. Kwok, P. Wang, L.S. Taylor, D. Ben-Amotz, Rapid classification of pharmaceutical ingredients with Raman spectroscopy using compressive detection strategy with PLS-DA multivariate filters, Journal of Pharmaceutical and Biomedical Analysis, 80 (2013) 63-68.

[154] M.R. de Almeida, D.N. Correa, W.F.C. Rocha, F.J.O. Scafi, R.J. Poppi, Discrimination between authentic and counterfeit banknotes using Raman spectroscopy and PLS-DA with uncertainty estimation, Microchemical Journal, 109 (2013) 170-177.

[155] M. Muratore, Raman spectroscopy and partial least squares analysis in discrimination of peripheral cells affected by Huntington's disease, Anal Chim Acta, 793 (2013) 1-10.

[156] V. Sikirzhytski, K. Virkler, I.K. Lednev, Discriminant Analysis of Raman Spectra for Body Fluid Identification for Forensic Purposes, Sensors, 10 (2010) 2869-2884.

[157] V.A. Sirotkin, Effect of dioxane on the structure and hydration-dehydration of alpha-chymotrypsin as measured by FTIR spectroscopy, Biochim Biophys Acta, 1750 (2005) 17-29.

[158] S. Yadav, N.G. Pinto, G.B. Kasting, Thermodynamics of water interaction with human stratum corneum I: measurement by isothermal calorimetry, J Pharm Sci, 96 (2007) 1585-1597.

[159] M. Takenouchi, H. Suzuki, H. Tagami, Hydration characteristics of pathologic stratum corneum--evaluation of bound water, J Invest Dermatol, 87 (1986) 574-576.

[160] M. Gniadecka, O. Faurskov Nielsen, D.H. Christensen, H.C. Wulf, Structure of water, proteins, and lipids in intact human skin, hair, and nail, J Invest Dermatol, 110 (1998) 393-398.

[161] P. Leary, F. Adar, R. Carlton, J. Reffner, F. Kang, R. Mueller, Hydration Studies of Pharmaceuticals Using IR and Raman Spectroscopy, Microscopy and Microanalysis, 13 (2007) 1694-1695.

[162] T. Richter, J.H. Müller, U.D. Schwarz, R. Wepf, R. Wiesendanger, Investigation of the swelling of human skin cells in liquid media by tapping mode scanning force microscopy, Applied Physics A: Materials Science & Processing, 72 (2001) S125-S128.

[163] R.R. Warner, K.J. Stone, Y.L. Boissy, Hydration disrupts human stratum corneum ultrastructure, J Invest Dermatol, 120 (2003) 275-284.

[164] J.A. Bouwstra, G.S. Gooris, J.A. van der Spek, S. Lavrijsen, W. Bras, The lipid and protein structure of mouse stratum corneum: a wide and small angle diffraction study, Biochim Biophys Acta, 1212 (1994) 183-192.

[165] T.J. McIntosh, M.E. Stewart, D.T. Downing, X-ray diffraction analysis of isolated skin lipids: reconstitution of intercellular lipid domains, Biochemistry, 35 (1996) 3649-3653.

[166] G.C. Charalambopoulou, T.A. Steriotis, T. Hauss, A.K. Stubos, N.K. Kanellopoulos, Structural alterations of fully hydrated human stratum corneum, Physica B: Condensed Matter, 350 (2004) E603-E606.

[167] J.A. Bouwstra, A. de Graaff, G.S. Gooris, J. Nijsse, J.W. Wiechers, A.C. van Aelst, Water distribution and related morphology in human stratum corneum at different hydration levels, J Invest Dermatol, 120 (2003) 750-758.

[168] M. Egawa, M. Oguri, T. Kuwahara, M. Takahashi, Effect of exposure of human skin to a dry environment, Skin Res Technol, 8 (2002) 212-218.

[169] M. Egawa, T. Kajikawa, Changes in the depth profile of water in the stratum corneum treated with water, Skin Res Technol, 15 (2009) 242-249.

[170] M. Wegener, R. Neubert, W. Rettig, S. Wartewig, Structure of stratum corneum lipids characterized by FT-Raman spectroscopy and DSC. III. Mixtures of ceramides and cholesterol, Chem Phys Lipids, 88 (1997) 73-82.

[171] S. Wartewig, R. Neubert, W. Rettig, K. Hesse, Structure of stratum corneum lipids characterized by FT-Raman spectroscopy and DSC. IV. Mixtures of ceramides and oleic acid, Chemistry and Physics of Lipids, 91 (1998) 145-152.

[172] J.N. Kraft, C.W. Lynde, Moisturizers: what they are and a practical approach to product selection, Skin Therapy Lett, 10 (2005) 1-8.

[173] C.W. Lynde, Moisturizers: what they are and how they work, Skin Therapy Lett, 6 (2001) 3-5.

[174] T. Pott, Moisturizers (hydratants) and Cosmétique. Entre mythe, réalité et controverse, Le MiDiFABs, 5 (2006) 21-43.

[175] A. Rawlings, C. Harding, A. Watkinson, J. Banks, C. Ackerman, R. Sabin, The effect of glycerol and humidity on desmosome degradation in stratum corneum, Arch Dermatol Res, 287 (1995) 457-464.

[176] I.R. Scott, Harding, C.R., Physiological effects of occlusion-filaggrin retention Proc Dermatol, (1993) 773.